3/02
4C

The Microcontroller
Beginner's Handbook
Second Edition

The Microcontroller
Beginner's Handbook
Second Edition

by Lawrence A. Duarte

PROMPT® PUBLICATIONS

International Standard Book Number: 0-7906-1153-8

Acquisitions Editor: Candace Hall, Loretta Yates
Editor: Natalie F. Harris
Assistant Editors: Pat Brady, Loretta Yates
Typesetting: Natalie Harris
Indexing: Natalie Harris
Cover and CD-ROM Design: Phil Velikan
Graphics: Lawrence Duarte
Graphics Conversion: Kelli Ternet, Terry Varvel, Natalie Harris, Christy Pierce
CD-ROM: Lawrence Duarte
Additional Materials: Courtesy of Adobe, Cypress Semiconductor, Gernsback Publications, Harris Semiconductor, Intel Corporation, Microchip Technology Incorporated, Motorola, Parallax, Inc., Scenix Semiconductor, Texas Instruments, Zilog.

PRINTED IN THE UNITED STATES OF AMERICA

9 8 7 6 5 4 3 2 1

Table of Contents

Dedication

For my Mother and Father

Acknowledgment

Special thanks to my good friend Chris Christner,
who spent many a long night reviewing and editing this book.

Preface

In the movie *Back to the Future, Part II*, the hero Marty, through a misadventure, is drenched with water. His futuristic jacket senses the water, turns on a fan and heater element, and dries the jacket while giving verbal progress reports to Marty. Is this the future? No, this is the present. With advances in technology — especially microcontrollers — the future is here today. Microcontrollers can be found everywhere: microwave ovens, coffee makers, telephones, cars, toys, TVs, clothes washer/dryer, etc. The list is endless.

This book will bring you information on how to understand, repair, or design a device incorporating a microcontroller. All elements of microcontroller use will be examined, including such industrial considerations as price vs. performance. Firmware will also be extensively analyzed and explained. A wide variety of third-party development tools, both hardware and software, will be seen. Emphasis is on new project design; however, once the reader has the knowledge to use a microcontroller, it will greatly enhance the ability to repair such devices.

We will start with the basics. All that is required is the knowledge of a first-year electronics course. References will be made to resistors, capacitors, RC constants, DC - AC power and other elementary terms. If the previous terms are Greek to you, then it might be advisable to review a beginning electronics book. If you know what they are but have a queasy feeling in your stomach as to how much you know about them, dive in! You have the basic knowledge, and this book will provide the rest. Finally, some readers will find many of the paragraphs in this book 'elementary,' and entire chapters may seem to be little more than review. Read on! There are many ideas and hints for even an experienced microcontroller user.

Microcontrollers by their very nature are digital devices. Some readers may have only experienced electronics in the analog realm. As we shall see, the digital-analog connection is very important to microcontrollers. We cannot use microcontrollers without understanding digital, or binary, manipulation. The field of binary operations is called Boolean Logic. Do not panic if these words are strange and unusual to you. We will cover them in great detail to provide you with the tools to use a microcontroller.

With such a huge number of applications for microcontrollers, the market for them is vast. Motorola alone sells 10 million a month of the 68HC11. Some other manufactures are Intel, Zilog, Toshiba, Sharp, and Microchip. While some comparisons of these different products will be made in this book, if we covered each product line in depth, this book would be several volumes long. Therefore, a very difficult choice was made to single out one particular line of microcontrollers for study. Emphasis will be on the Microchip PIC® products. However, do not let this textual preference lead you to think that Microchip PIC® products are the only ones to use. To paraphrase a sage, I have not seen a microcontroller that I did not like.

Theory is fine, but doing is better. After you gain a firm knowledge of hardware and firmware, you will be able to put it to use by building five projects. They will progress from simple to very sophisticated. All relevant schematics, board layouts, and firmware will be provided and explained. Don't let this intimidate you. Any hobbyist should be able to complete these projects. When done, you will be competent to design, write, and build a new project utilizing a microcontroller.

—Lawrence A. Duarte

CHAPTER 1
What is a Microcontroller?

What are these small pieces of silicon that are changing the way we live on a daily basis? As with any *integrated circuit* (IC), they are a collection of thousands of *transistors* that fit in an area approximately 1/10" square. The silicon chips are packaged in plastic or ceramic carriers, and resemble any other IC on the market. They can have 16, 18, 24, or any number of pins that suit them. Some are *dual in-line packages* (DIP); others are surface mount; and some even come with windows. *Figure 1-1* shows the physical appearance of several different microcontrollers.

The internal workings are similar to standard TTL- or CMOS-type chips. Indeed, thousands of the logic gates used in these other chips are also contained in a microcontroller. A *programmable gate array* (PGA) is like a microcontroller in that it can be programmed to reconfigure internal logic for a variety of tasks. A microprocessor (the heart of your average PC) is often confused with a microcontroller. Rightly so since a microprocessor has an internal instruction set and acts upon programming. What, then, separates a microcontroller from a logic chip, PGA, or a microprocessor? We will answer that question shortly.

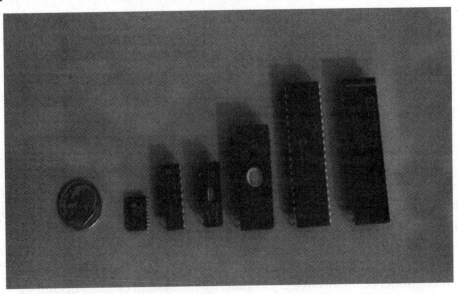

Figure 1-1. *Microcontrollers come in all shapes and sizes.*
Some are the size of half a dime.

Finally, microcontrollers themselves are wildly different from each other. One may contain ten or a hundred times the transistors of another. Some have analog elements built in. Others have nonvolatile (or permanent) memory. Many microcontrollers have been modified to fit a particular purpose such as home automation, communications, or computer peripherals. The internal working size can be 4, 8, 12, 16 bits (we will explain bits momentarily). Can we find common ground to catalog these ICs into a unique group?

A Definition

Microcontrollers must contain at least two primary components—*random access memory* (RAM), and an instruction set. RAM is a type of internal logic unit that stores information temporarily. RAM contents disappear when the power is turned off. While RAM is used to hold any kind of data, some RAM is specialized, referred to as *registers*. The instruction set is a list of all commands and their corresponding functions. During operation, the microcontroller will step through a program (the *firmware*). Each valid instruction of the program is a command to the microcontroller to do a specific job. It is the instruction set and the matching internal hardware that differentiate one microcontroller from another.

Most microcontrollers also contain *read-only memory* (ROM), *programmable read-only memory* (PROM), or *erasable programmable read-only memory* (EPROM). All of these memories are permanent: they retain what is programmed into them even during loss of power. They are used to store the firmware that tells the microcontroller how to operate. They are also used to store permanent lookup tables. Often these memories do not reside in the microcontroller; instead, they are contained in external ICs, and the instructions are fetched as the microcontroller runs. This enables quick and low-cost updates to the firmware by replacing the ROM.

Where would a microcontroller be without some way of communicating with the outside world? This job is left to *input/output* (I/O) port pins. The number of I/O pins per controller varies greatly, plus each I/O pin can be programmed as an input or output (or can even switch during the running of a program!). The load (current draw) that each pin can drive is usually low. If the output is expected to be a heavy load, then it is essential to use a driver chip or transistor buffer.

Most microcontrollers contain circuitry to generate the system clock. This square wave is the heartbeat of the microcontroller, and all operations are synchronized to it. Obviously, it controls the speed at which the microcontroller functions. All that is needed to complete the clock circuit would be the crystal or RC components. We can, therefore, precisely select the operating speed critical to many applications.

Now we have a wide choice of possible supplemental modules. The details of these will be saved for later in the book. A great number of them have an internal reset (starts the microcontroller fresh on powering up) and watchdog (resets the controller after a problem occurs) circuits. Many microcontrollers also contain nonvolatile memory: analog-to-digital, digital-to-analog, serial port, I^2C bus, timers, and interrupts.

To summarize, a microcontroller contains (in one chip) two or more of the following elements in order of importance:

A. Instruction set
B. RAM
C. ROM, PROM, or EPROM
D. I/O ports
E. Clock generator
F. Reset function
G. Watchdog timer
H. Serial port
I. Interrupts
J. Timers
K. Analog-to-digital convertors
L. Digital-to-analog convertors
M. I^2C bus

Low Cost

In terms of capability for cost, the closest comparisons to the microcontroller would be the microprocessor and the *application specific integrated circuit* (ASIC). Both of these alternatives are vastly more expensive than the microcontroller, and in the case of an ASIC, the engineering fees (creating the custom IC die) are prohibitive. Even discrete logic (multiple independent ICs) is often more expensive than a single microcontroller. Cost is an important factor to consider when adding intelligence to an appliance like a coffee machine; after all, adding thirty dollar's worth of electronics to a twenty dollar appliance is not commercially feasible.

For example, let's look at the pricing of a Microchip PIC® 16C57. The one time-program-mable (OTP) versions sell for around $7.50 each. At 100 pieces, the price drops to $4.50 each. Purchase several hundred PICs® from a distributor, and the price can fall to $3.50 each. This is a lot of processing power for very little money. It is true that an ASIC manufactured in large quantities can fall below this per-piece price; however, the average $50,000 setup fee has to be factored in.

Some other examples of controller cost in single quantity:

8031	40 pins, 128 bytes RAM, I/O	**$3.25**
80C31	CMOS (low power), 40 pins, 128 bytes RAM, I/O	**$5.49**
DS80C320	High speed (25 MHz) version of 80C31	**$12.95**
8049	8-bit microcontroller (one of the oldest!)	**$0.95**
8052AHBASIC	Microcontroller with internal Basic	**$24.95**
MC68HC11A1P	Microcontroller with internal A/D	**$10.95**
Z80H	40 pins, 8 MHz speed	**$2.95**

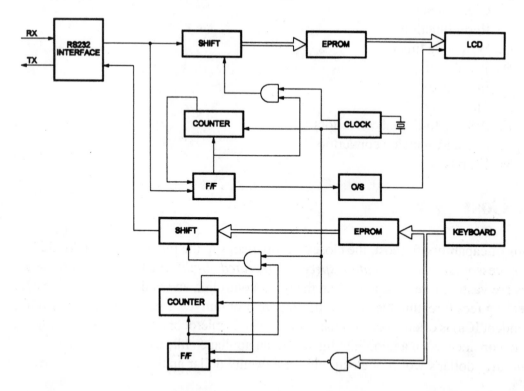

Figure 1-2. *Block diagram for a discrete logic hand-held RS232 Terminal.*
Parts for device cost approximately $20.14.

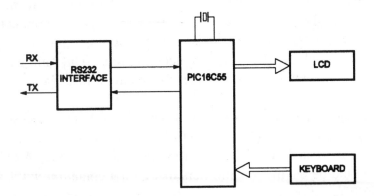

Figure 1-3. *The same RS232 Terminal as in Figure 1-2, but utilizing a microcontroller. Total cost is around $11.59.*

Obviously, the price varies quite a lot depending on internal capabilities and speed. But how does it compare against discrete logic?

Figure 1-2 shows a block diagram for a discrete logic hand-held RS232 terminal. It contains 12 chips and has an approximate cost of $20.14. *Figure 1-3* shows the same application using a microcontroller. It uses three chips at a cost of $11.59. Lower costs are also achieved by reducing the size of the *printed circuit board* (PCB), requiring much less time to build one. When all costs are factored in, the controller version will be less than half the price of the discrete unit. Also, be aware that *Figure 1-2* shows a bare-bones circuit without parity or macro-command capabilities. The microcontroller can do all of this plus more for no extra cost. This is a compelling argument for using microcontrollers.

Powerful

The processing power of a microcontroller is directly related to two parameters: clock speed and the instruction set. The controller operates synchronously with the clock; however, it does not necessarily operate *as fast* as the clock. An example would be the PIC® microcontroller in *Figure 1-3*. If it used a 20-MHz crystal, the clock would be said to be 20 MHz. But this controller takes four clock cycles to execute one instruction. The PIC® therefore operates at 1/4 the clock rate, or 5 MHz. Another way to measure performance is by the number of instructions executed per second. A 5-MHz rate is referred to as a *5-million-instructions-per-second* (MIPS) rate, and is the standard way to compare speed among microcontrollers.

For example, the venerable 80C31 takes twelve clock cycles to perform one instruction; so, given a 12-MHz clock, the 80C31 gives you 1 MIPS. Dallas Semiconductor has issued a high-speed version of the 80C31, the DS80C320. It is a pin-compatible replacement that decreases the number of clock cycles per instruction to four. This chip would instantly increase the MIPS to three with no other changes to the circuit board!

The instruction set also plays an important roll when considering processing power. If a controller is lacking a commonly-used instruction, then a work-around or software subroutine used to compensate can decease the performance by a magnitude. A common example of this problem would be the microcontrollers that do not contain a *multiply* or *divide* instruction. If your application requires a multiply operation, then the only solution is a software subroutine. This could slow your program from microseconds (using the built-in multiply) to milliseconds (using a work-around). It is true that most microcontroller applications are not numerically intensive, but should they be—compare the instruction set with your requirements before selecting the microcontroller.

Having examined the fundamentals of possessing power, the beginning user of microcontrollers is probably lost in MIPS, clocks cycles and instructions. To get a better grip on the issues, let's look at a few real-life examples. By comparing *Figure 1-2* to *Figure 1-3*, you saw how the same performance could be achieved in a much more efficient design. As a bonus, the greatly-enhanced intelligence of the microcontroller design enables it to use such extra functions as self diagnostics, macros per keystroke (a series of keystrokes), and programmable keys. Another example would be fish-finders, based on sonar signal transmission/reception and a graphic LCD output of the results. The only major electronic component is a standard microcontroller! One of the projects in this book is an indoor/outdoor thermometer gauge with a video output. While there are ICs exclusively made to generate a video image, this project will create a NTSC signal, plus create the information content of the signal and read temperature sensors, all from one standard 18-pin microcontroller! This shows the raw processing power of which microcontrollers are capable.

Flexible

An important element of any electronic design is the ability to update or correct the design after the product is finished. With microcontrollers, you have that ability in spades. This flexibility is seen in two respects: firmware and hardware. The firmware controls how the microcontroller functions, which in turn controls the entire application. Supporting ICs, discrete components, power supplies and PCBs may cost more than ten times the cost of the microcontroller. By placing a socket where the controller is normally soldered in place, you can then replace the microcontroller at will. A three dollar swap-out of the controller will often save a one-hundred dollar device.

The ability to change the firmware easily, quickly and inexpensively has many advantages, as shown in the following examples. A customer wanted a board that would automatically change the volume output of speakers based on the ambient noise in the room. He had to have a finished product quickly, but there were questions as to how the ambient noise would be received by the unit. The circuit was sound, so a PCB was created with the understanding that any corrections could be made later in the microcontroller. Ten beta units were prepared and sent out to end users for testing. After receiving feedback, the firmware was modified, new microcontrollers were cut, the units in the field were updated, and full production of the board began. *Figure 1-4* shows the production board.

Another example shows flexibility in a finished unit. The device in question controlled the operation of a projector lift. Based on button pushes, the lift would move the projector to one of two preset positions (not including the top). A customer desired to buy the lift but needed three preset positions. With nothing more that a simple change to the firmware, the

Figure 1-4. A production microcontroller product. This board will automatically adjust the volume of the speaker output based upon ambient noise level.

product the customer wanted was created (and the customer was charged $500 for one hour of work!). This also illustrates a maxim found in numerous microcontroller designs: socket the microcontroller! You never know when you may have to change or update the firmware.

A less frequently used function is the ability to change the hardware configuration of the microcontroller. In most designs, there will be I/O pins left over. It is a good practice to design the PCB with pads going to them with pull-up resistors on the trace. Should the need arise, you have extra inputs or outputs ready to go. There is also the possibility of changing a currently designated pin from an output (drives LED) to an input (reads button status). Sometimes, installing a resistor on the board is required. All of these hardware changes require firmware changes as well. The power and flexibility of microcontrollers has been a blessing to many such applications.

RISC

In closing this discussion on the nature of microcontrollers, we should examine the difference between *reduce instruction set computer* (RISC) and *complex instruction set computer* (CISC). The current most popular microcontrollers are of the CISC type. RISC controllers, however, have come on strong in the last few years and could easily eclipse the CISC. RISC controllers have traded a smaller instruction set for quicker speed and faster execution. The technical reasons for this tradeoff are complex and not suitable for this introduction, but here's a functional comparison between the two:

80C31	111 Instructions	12 MHz	1 MIPS
68HC11	62 Instructions	8 MHz	2 MIPS
TMP87C800N/F	129 Instructions	8 MHz	2 MIPS
PIC16C57	33 Instructions	20 MHz	5 MIPS
Z86C06	47 Instructions	12 MHz	6 MIPS

In the microcontroller field, there has been quite an argument between the advocates of RISC vs. CISC. In fact, the dividing line between the two architectures seems to decrease every day. The standard 80C31 has a MIPS of only 1, but there are such current high-speed versions as the DS80C320 (10 MIPS) and MCS251 (5 MIPS), with no loss in the instruction set. Many RISC controllers have added numerous new instructions, earning the oxymoron of *complex RISC*.

You have now seen a wide diversity in the performance and capabilities of today's microcontrollers. Careful thought should be given before the beginning of any design phase as to which microcontroller suits the job best. Many times, a number of different controllers will work equally well. The deciding factor, then, is often practical the practical side: which costs less, and do you have the tools (programmer, assembler) to work on the microcontroller?

CHAPTER 2
Binary and Boolean Logic

In the Chapter 1, you saw the numerous advantages behind microcontrollers. As with anything in life, there is a price to pay when using them. It is nearly impossible to write the firmware without knowledge of binary manipulation and Boolean logic. Even veteran programmers of microcomputers can be novices at this because they program in "high level" languages such as Visual Basic or Clipper (a database language). At this level, entire programs can be written with no referrals to I/O ports or Boolean logic.

The standard dictionary definition of binary is *characterized by or consisting of two parts or components; twofold.* This applies to computers because the basic unit of operation is a bit which has only two possible valid states: high or low. Boolean logic is name after George Boole (1815-64), who was a pioneer in the field of logic. Again, the dictionary tells us that Boolean is *of or relating to a logical combinatorial system treating variables, such as propositions and computer logic elements, through operators.* You shall see the very operators George Boole used, AND, OR, NOT, and others, are still used today.

It would be disingenuous to say that this chapter is easy. In fact, to those readers fresh to the concept of binary manipulation, this chapter may be the most difficult one in the book. The only foundation necessary is first-year algebra and an introduction to different number system bases. The best way to proceed is to read this chapter and understand it as best as you can. Then, read the rest of the book, as there will be plenty of examples illustrating the ideas further. Finally, read this chapter again once you've read through the entire book, and it will be much clearer.

Ones and Zeros

The basic unit in computers, whether they be mainframes, microcomputers, or microcontrollers, is the bit (a rather unpretentious word for something which has changed the world). A bit has only two valid states: one or zero. It is quite common to refer to these two states in three different ways. One can also be referred to as *high*, or in a 5-volt circuit *+5vdc*. A zero is *low*, or *0vdc*. This can be confusing to the novice. *Figure 2-1* illustrates the connection. This graphic box will be included on pages that discuss logic functions. The voltage reference is a clue as to how to connect the theoretical logic to actual use in the electronic circuit.

ONE (1) = HIGH = +5VDC

ZERO (0) = LOW = 0VDC

5 VOLT SYSTEM

Figure 2-1. *Equivalent terms in the microcontroller field.*

Obviously, by itself, one bit with two possible states is not of much value (beyond representing a button press *on* or *off*). However, by stringing the bits together, you get progressively longer logic units. Each bit added increases what the logic unit can hold by twofold. *Figure 2-2* shows this progression. *Figure 2-2* also shows the common terms for four bits (a *nibble*), and eight bits (a *byte*). You may ask why these lengths deserve special terms. Many functions are performed that use nibbles; you shall see some in this chapter and more later in the book. There are even microcontrollers that are based solely on the nibble, such as the Toshiba TLCS42 series. Microcontrollers based on bytes (eight bits) are much more common, occupying 90% of the controller market. There is nothing intrinsically magical about these numbers. The PIC® family of controllers includes 12 bits in the instructions, some bits for the command, and other bits for addressing or options. When a PIC® performs math, it operates on bytes, and is therefore considered an 8-bit processor.

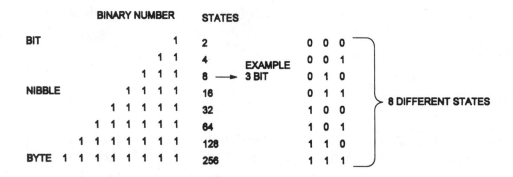

Figure 2-2. *Here you can see the value that can be held in an increasing binary number; one bit can hold two states, one nibble can hold sixteen states, and one byte can hold 256 states.*

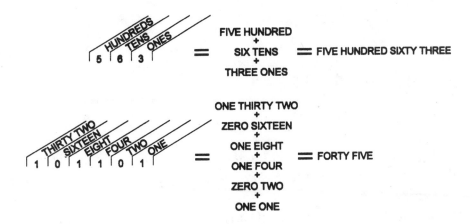

Figure 2-3. *The total value of a number, whether decimal or binary, can be derived by multiplying the column value by the number in the column and adding the results.*

Though we will not be discussing them much, there are microcontrollers whose standard logic unit is 16 or 32 bits. These high-end controllers are very powerful and somewhat expensive. The term for 16 bits (two bytes) is *word*. Two words (32 bits) is a *double word*. These controllers can handle 65,536 and 4,294,967,296, respectively, in their basic unit.

Binary Arithmetic

Learning binary arithmetic is not as hard as it might sound. The core of a good learning foundation is the knowledge that binary numbers use a *base two* number system. The every-day number system is *base ten* (decimal). This means you count ten digits (0, 1, 2, 3, 4, 5, 6, 7, 8 ,9) then roll over to the 10s column. You then count ten 10s and roll over to the hundreds. Similarly, in base two, you count two digits (0, 1) before rolling over to the "10s" column. Of course, that would actually be the "2s" column in base two. It is important to remember that in base, two there are only two valid digits: 0 and 1. The "roll over" reference is properly called a *carry*, something you use in everyday addition. When you add 8 to 16, there is a carry from the 1 column to the 10 column, and the result is 24. Let's add 1 to 1 in binary. This produces a value that cannot be contained in the 1 column of a base two number. Just as in base ten, you have a carry. Therefore, 1 plus 1 is 1 0 (not 10, but one zero). *Figure 2-3* shows a comparison of column values between a base ten number and a base two number.

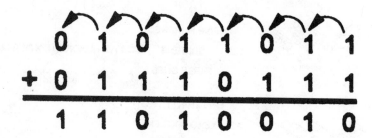

Figure 2-4. In binary addition, there is no 2, so 1 + 1 produces 0
with a carry into the next column.

Binary addition usually involves a string of carries. An example is seen in *Figure 2-4*. Each column of the two 8-bit numbers produced a carry (except the last). 8-bit additions are quite common in microcontrollers. If the last column had produced a carry, then a special flag would be set high. You will learn more about this later. What is the quickest way to convert these numbers to decimal? A lookup chart is quite nice, as in *Table 2-1*. By using this chart, you can see that the operands in *Figure 2-4* are 91 and 119, giving the result of 210. Most scientific calculators do conversions between binary and decimal. To do it by hand, simply add each column's value (if it contains a 1) to produce the value of the entire number. Look at *Figure 2-3* to see how this is done.

Binary subtraction is very similar to addition, except that instead of a carry, you have a *borrow*. *Figure 2-5* shows typical binary subtraction. It may be little harder to grasp that the borrow is 2. The reason for this is that each number in the column to the left is worth two of the column immediately to the right of it. However, you may say, there is no number 2 in binary! That's true, and no number 2 will ever appear since you will only borrow when you are about to subtract a 1 from the borrow, producing a 1 as the result. If the second operand is larger than the first operand, then the result is negative. In most microcontrollers, the carry

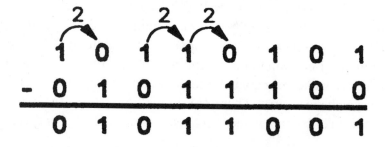

Figure 2-5. In binary subtraction, a borrow is value 2.

Decimal	Binary	Decimal	Binary	Decimal	Binary	Decimal	Binary
0	O0000000	64	O1000000	128	10000000	192	11000000
1	O0000001	65	O1000001	129	10000001	193	11000001
2	O0000010	66	O1000010	130	10000010	194	11000010
3	O0000011	67	O1000011	131	10000011	195	11000011
4	O0000100	68	O1000100	132	10000100	196	11000100
5	O0000101	69	O1000101	133	10000101	197	11000101
6	O0000110	70	O1000110	134	10000110	198	11000110
7	O0000111	71	O1000111	135	10000111	199	11000111
8	O0001000	72	O1001000	136	10001000	200	11001000
9	O0001001	73	O1001001	137	10001001	201	11001001
10	O0001010	74	O1001010	138	10001010	202	11001010
11	O0001011	75	O1001011	139	10001011	203	11001011
12	O0001100	76	O1001100	140	10001100	204	11001100
13	O0001101	77	O1001101	141	10001101	205	11001101
14	O0001110	78	O1001110	142	10001110	206	11001110
15	O0001111	79	O1001111	143	10001111	207	11001111
16	O0010000	80	O1010000	144	10010000	208	11010000
17	O0010001	81	O1010001	145	10010001	209	11010001
18	O0010010	82	O1010010	146	10010010	210	11010010
19	O0010011	83	O1010011	147	10010011	211	11010011
20	O0010100	84	O1010100	148	10010100	212	11010100
21	O0010101	85	O1010101	149	10010101	213	11010101
22	O0010110	86	O1010110	150	10010110	214	11010110
23	O0010111	87	O1010111	151	10010111	215	11010111
24	O0011000	88	O1011000	152	10011000	216	11011000
25	O0011001	89	O1011001	153	10011001	217	11011001
26	O0011010	90	O1011010	154	10011010	218	11011010
27	O0011011	91	O1011011	155	10011011	219	11011011
28	O0011100	92	O1011100	156	10011100	220	11011100
29	O0011101	93	O1011101	157	10011101	221	11011101
30	O0011110	94	O1011110	158	10011110	222	11011110
31	O0011111	95	O1011111	159	10011111	223	11011111

Continued...

32	O0100000	96	O1100000	160	10100000	224	11100000
33	O0100001	97	O1100001	161	10100001	225	11100001
34	O0100010	98	O1100010	162	10100010	226	11100010
35	O0100011	99	O1100011	163	10100011	227	11100011
36	O0100100	100	O1100100	164	10100100	228	11100100
37	O0100101	101	O1100101	165	10100101	229	11100101
38	O0100110	102	O1100110	166	10100110	230	11100110
39	O0100111	103	O1100111	167	10100111	231	11100111
40	O0101000	104	O1101000	168	10101000	232	11101000
41	O0101001	105	O1101001	169	10101001	233	11101001
42	O0101010	106	O1101010	170	10101010	234	11101010
43	O0101011	107	O1101011	171	10101011	235	11101011
44	O0101100	108	O1101100	172	10101100	236	11101100
45	O0101101	109	O1101101	173	10101101	237	11101101
46	O0101110	110	O1101110	174	10101110	238	11101110
47	O0101111	111	O1101111	175	10101111	239	11101111
48	O0110000	112	O1110000	176	10110000	240	11110000
49	O0110001	113	O1110001	177	10110001	241	11110001
50	O0110010	114	O1110010	178	10110010	242	11110010
51	O0110011	115	O1110011	179	10110011	243	11110011
52	O0110100	116	O1110100	180	10110100	244	11110100
53	O0110101	117	O1110101	181	10110101	245	11110101
54	O0110110	118	O1110110	182	10110110	246	11110110
55	O0110111	119	O1110111	183	10110111	247	11110111
56	O0111000	120	O1111000	184	10111000	248	11111000
57	O0111001	121	O1111001	185	10111001	249	11111001
58	O0111010	122	O1111010	186	10111010	250	11111010
59	O0111011	123	O1111011	187	10111011	251	11111011
60	O0111100	124	O1111100	188	10111100	252	11111100
61	O0111101	125	O1111101	189	10111101	253	11111101
62	O0111110	126	O1111110	190	10111110	254	11111110
63	O0111111	127	O1111111	191	10111111	255	11111111

Table 2-1. *A binary-to-decimal conversion table for one byte (256 numbers).*
(Continued from Page 17.)

flag is used to indicate a borrow generated if the second operand is larger than the first. It might seem confusing to use carry to represent borrow, but in programs where you must see if there was a carry or borrow, you check the flag immediately after the operation so you can see if the carry was a result of an addition or subtraction.

It is interesting to note how a subtraction is performed. In the controller, subtraction is done by taking the two's complement (see the NOT function) of the number being subtracted, then performing an addition to the first operand. (Disregard the last carry. If there is no carry, then the answer is negative, and the two's complement is taken of that answer). This is the same as subtraction; try it yourself. While it may seem complicated, the internal circuit for addition is already present, thereby making this algorithm easy to implement.

Hex Arithmetic

Since we are talking about different number systems, let's look at the hexadecimal (hex) number system. This is a base 16 number system. In the microcontroller field, as well as in computers in general, hex numbers have proven to be very convenient. One of the primary reasons is because we often talk about two-byte numbers (addressing, data, I/O ports), and to write such long strings of numbers as 1011011100111100 in programs or documentation would be tedious and error prone. With hex numbering, you can write the above number as B73C. Another advantage is that the conversion from binary to hexadecimal is made in seconds without any lookup tables or a calculator. You shall see how shortly.

Hex digits, being base-16 number systems, run from 0 to 15. Of course, you can't use 10,11,12,13,14,15 to represent a single digit (that would be confusing). So, you use the next best thing: the alphabet! Hex digits are 0, 1, 2, 3, 4, 5, 6, 7, 8, 9, A, B, C, D, E, F. Each column of numbers is worth 16 of the column to the right. An example would be the number A1: A is the value 10; each digit in this column is worth 16 of the column to the right (ones); thus 10 x 16 is 160 plus the 1 in the ones column, producing a grand total of 161. That's the conversion to decimal, but how about the quick and easy conversion between hex and binary?

As you now know, each column in hex will hold a value of 16 (0 to F). A nibble (4 bits) in binary will also hold exactly a value of 16 (0000 to 1111). This permits a direct replacement of each hex number by its binary equivalent, and vice versa. The most difficult part is memorizing the first 16 conversions, seen in *Table 2-2*. This chart quickly becomes second nature to veterans of microcontrollers. Now the conversion of much larger numbers is as simple as replacing each hex digit with four bits, or four bits with a hex digit. *Figure 2-6* shows this translation in action.

Most compilers and assemblers (tools programmers use to write code) will readily accept decimal, binary or hex numbers. However, there must be something to tell the compiler which number system it is working with; after all, 10 could be hex (16), binary (2), or decimal (10). This notation varies from compiler to compiler, but a common method is to add *h* to the end of a hex entry, nothing for a decimal entry, and *b* to the end of a binary entry. For example, 18h is hex 18, 1011b is binary 1011, and 150 would be decimal. Unlike binary or decimal numbers, hex numbers can start with a letter. To avoid confusing the compiler, a 0 is always added to the start of hex number when it begins with a letter. Thus, A7h becomes 0A7h.

Binary	Hex
0 0 0 0	0
0 0 0 1	1
0 0 1 0	2
0 0 1 1	3
0 1 0 0	4
0 1 0 1	5
0 1 1 0	6
0 1 1 1	7
1 0 0 0	8
1 0 0 1	9
1 0 1 0	A
1 0 1 1	B
1 1 0 0	C
1 1 0 1	D
1 1 1 0	E
1 1 1 1	F

Table 2-2. A binary-to-hex conversion table for the first sixteen numbers.

BCD Arithmetic

We saw that the conversion between hex and binary is easy because of the direct replacement of one hex digit with four binary bits. Is there a similar method for conversion to decimal, you might ask? Yes, and that's where binary-coded decimal (BCD) comes in.

BCD is based on the nibble, as with hex conversion, but because you are dealing with decimal, only the first ten values of the nibble are used. This gives you a direct correspondence, seen in *Table 2-3*. The remaining numbers of the nibble (1010 - 1111) are there, but not valid. By using the chart in *Table 2-3* and applying the same conversion technique used in hex to binary, you can quickly change 3453 to 0011010001010011 BCD.

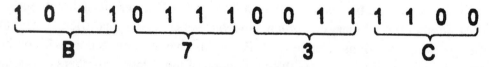

Figure 2-6. By using Table 2-2, you can quickly translate a long binary number to a hex.

What advantages are there to using BCD? In computers, all numbers must be represented in binary. Before microprocessors and microcontrollers, it was a complex process to convert straight binary to decimal. BCD enabled simplified translation to decimal, for both input and output to the real world. An example of this heritage is the 7446 or 7447 chips, as seen in *Figure 2-7*. They receive four-bit BCD code and directly drive a seven-segment LED display. Another example would be the 7442 BCD to 10-line decimal decoder IC. By applying BCD to this chip, one of ten outputs will be driven high. Even microcontrollers show respect for BCD by having special commands present. The 8031 family has the *decimal-adjust accumulator for addition* (DAA) command, which will adjust an internal register after the addition of two BCD numbers, producing a valid BCD number.

BCD	Decimal
0 0 0 0	0
0 0 0 1	1
0 0 1 0	2
0 0 1 1	3
0 1 0 0	4
0 1 0 1	5
0 1 1 0	6
0 1 1 1	7
1 0 0 0	8
1 0 0 1	9

Table 2-3. *A BCD-to-decimal conversion table.*

BCD is of limited use today. Fast internal subroutines in firmware can convert between decimal and binary without the need of BCD. BCD is little more than a waste of memory space. The maximum value that BCD can hold in eight bits is 99, while full binary can hold 256 in eight bits. This is even more extreme when dealing with sixteen bits; maximum value in BCD is 9,999 while in binary it is 65,535! However, keep BCD in mind. There will come a time when you might be dealing with old devices requiring BCD, or you choose to use old support chips to drive a LED display.

The NOT Function

Up to now, you have being dealing with actual numbers and the addition, subtraction, and conversion of those numbers. The following sections deal with logic functions, which do not use traditional mathematics. An example would be 1 AND 1 equals 1; this is a true statement because you are dealing with logic and not arithmetic. *Figure 2-1* again reminds you of the relationship between firmware verbiage and hardware reality. Another synonym could have

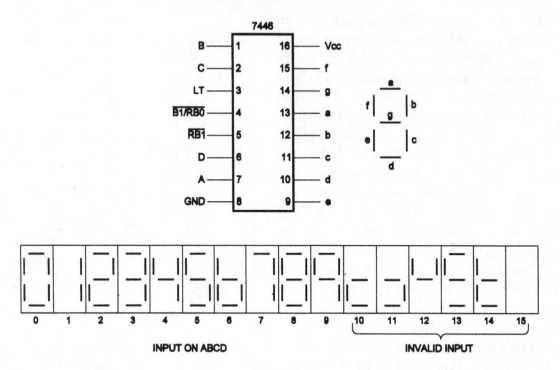

Figure 2-7. *The 7446 IC will take a BCD input and drive a seven-segment LED display.*

been included in *Figure 2-1* which relates to logic: 1 is True and 0 is False. The other terms are much more commonly used and will be seen in this book.

The NOT function is the simplest of logic functions; it is merely the opposite of a binary bit. NOT 1 is 0. When the NOT function is performed on a byte, it becomes the one's complement of that number. A sample assembler code for this is COMF, one's complement command in a PIC® microcontroller. Parallax instruction set calls the complement command NOT even though it operates on an entire byte. NOT 10010110 would become 01101001. The two's complement is found by adding one to the one's complement. An example of two's complement would be 001 becomes 110 + 1 or 111. You have seen a practical use of two's complement in the previous section on subtraction.

A *logic table* is used to display inputs and outputs of logic functions. The table is divided into two major sections; the left side is inputs, and the right side is outputs. Each column on the right or left represents an independent input or output. *Figure 2-8* shows the logic table to the NOT function and examples of one's complement as well as two's complement. While

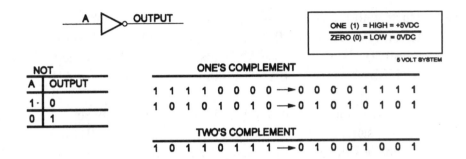

Figure 2-8. *The output of a NOT function is the opposite of the input.*

the logic table for NOT is small and easy, some logic functions are quite complicated, and the table then becomes a very useful tool. You also can use logic tables to help plan inputs and outputs for a microcontroller.

The AND Function

The function AND operates the same as its meaning in the English language. The statements *glass is clear*, *water can freeze* are true. Therefore, the statement *glass is clear AND water can freeze* is true as well. The statements *white is black*, *computers are free* are false. The compound statement *glass is clear AND white is black* consist of one true fact and one false fact; the AND junction makes the whole statement false. Of course, both false facts, *white is black AND computers are free*, make a false statement as well. You can see that logic functions are intuitive at an English language level.

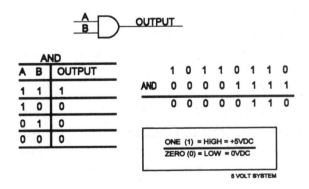

Figure 2-9. *A logic table for an AND function; sample bytewise AND.*

Replace the true facts with 1, and the false facts with 0. You can immediately see 1 AND 1 = 1 (true AND true is true), 1 AND 0 = 0 (true AND false is false), 0 AND 1 = 0 (false AND true is false), and lastly 0 AND 0 = 0 (false AND false is false). This is exactly what the AND function does on a bit-by-bit level. *Figure 2-9* shows the logic table for AND.

Most microcontrollers do not perform AND on a bit, but rather between bytes. To do this, the controller takes the same position bit in the bytes, executes an AND on them, and places the answer in the same position bit in the result. For example, take 10110110 and 00001111. The bits to the farthest right (called the least significant bit) are 0 and 1. After you AND 0 and 1, you put the answer 0 in the least significant bit of the result. The rest of the AND between these two bytes can be seen in *Figure 2-9*.

What good is the AND function? The uses vary, and you shall see many examples in the sample projects in this book. One use is called *masking*. Let's say one byte contains the status of eight switches; each bit represents a switch opened (1) or closed (0). The byte is currently 11000101. You want to know the if the seventh switch is opened or closed. The mask byte would be 01000000. 11000101 AND 01000000 is 01000000. As you can see, this action strips off unnecessary bits. After applying the mask byte, it is easy for the controller to tell if the seventh switch is open (result greater than zero) or closed (result is zero). Another example is to break a byte into two nibbles. This can be seen in *Figure 2-9*. The masking byte would be 00001111 or 11110000.

The OR Function

The OR function is like the AND function, the primary difference being that you need only one true statement for the whole statement to be true. Using the previously defined facts, the statement *water can freeze OR white is black* is true because one of the two facts is true. Both *glass is clear OR water can freeze* and *computers are free OR glass is clear* are true statements as well. The only false combination statement would be *white is black OR computers are free*.

In binary, the statements would be 1 OR 1 = 1, 1 OR 0 = 1, 0 OR 1 = 1, and 0 OR 0 = 0. *Figure 2-10* shows the OR logic table and examples. Most microcontrollers perform OR on bytes, not bits. As with the AND function, the result of the byte-wise OR is the sequential OR of each bit between the two bytes. Again, refer to *Figure 2-10* for an example of the OR function on two bytes.

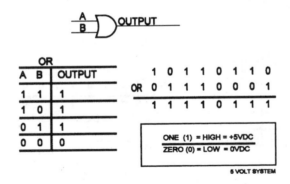

Figure 2-10. *A logic table for an OR function; sample bytewise OR.*

During a discussion of logic operators, there is often a reference to the answer or result byte. There are two locations where the result is placed after the operation, depending on the type of microcontroller you are using. With PIC® processors, the result is stored back into the location of one of the original byte operands. An example in Parallax assemble language is *OR 9,#23*. We will discuss at great length assembly language, but for now it is sufficient to know you are performing an OR function between the value in register 9 and the number 23. The result is stored back into register 9, replacing the original value. The other method used in controllers like the 80C31 is to retain both operand values and store the answer in a special register called the *accumulator*. There is a price to pay for this, as you must first copy one of the operands into the accumulator before the OR operation.

OR is used much less frequently than the AND function. It is sometimes used to set unnecessary bits high (10100000 OR 00001111 = 10101111), or to mix status bytes from two different locations representing the same switches. OR can also be found in unusual mathematics or graphic operations.

The XOR Function

The exclusive-OR (XOR) function has no English equivalent. Its operation is best learned by memorizing the simple logic table shown in *Figure 2-11*. Another way to remember is that if the bits are the same, then the result is false; if the bits are different, then the result is true. Once again, this operation is used in bytes in most microcontrollers. *Figure 2-11* shows a bytewise XOR.

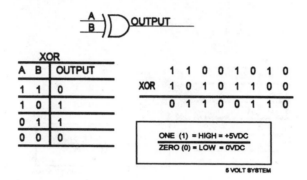

Figure 2-11. *A logic table for an XOR function; sample bytewise XOR.*

XOR, while seemingly bizarre, has greater use than the OR operation. XOR is used for such applications as parity generation, byte comparison, graphics, and math routines. Let's demonstrate one example: byte comparison. Suppose you are looking for 10110010. If you XOR the byte in question with the number you are looking for, the result will always be 0 if equal, and greater than 0 if not equal. You can see this by 10110010 XOR 10110010 = 00000000, while one possible non-match would be 10110010 XOR 10101110 = 0001100.

At this point, you may be asking, why use the XOR in byte comparison? Would not the microcontroller have to compare the resultant byte to 0 to see if the first two bytes were equal? These are very astute questions that have informative answers. The internal architecture of microcontrollers differ greatly from each other. Some controllers have a direct compare command of two bytes, and others can only do it in a convoluted way. Sometimes, an XOR is quicker than the internal compare operations, or provides more options to the programmer. You don't have to compare the result of XOR to 0 because of the 0 flag. In most microcontrollers, this is a special bit which immediately reports if the previous operation (such as XOR) has produced a 0 result. There are instructions that then enable you to perform jump, skip, or perform other functions based on the status of the 0 flag.

The Rotate Function

Rotate is not a logic or arithmetic function, but can roughly be grouped with the other operators. This function comes in two flavors: rotate right and rotate left, often with the option of using the carry flag. *Figure 2-12* shows the four different possibilities. The PIC® line can only perform rotate with carry, while the 80C31 has rotate with or without carry.

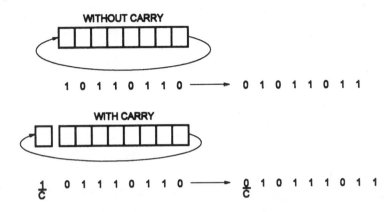

Figure 2-12. *The rotate function is like a bucket brigade;*
the bits are shifted left or right by one.

While rotate is not a true mathematical operation, it can be used for a simple multiply-by-two or divide-by-two operation. Recall that each bit in a byte is two times the value of the bit on the right. By shifting the whole byte to the left (to the most significant bit), the byte has been, in essence, multiplied by 2. Let's look at an example: 00110011 (51) rotate left = 01100110 (102) rotate left = 11001100 (204). *Divide by 2* is a rotate right. 10110000 (176) rotate right = 01011000 (88) rotate right = 00101100 (44). Obviously, rotate cannot handle fractions or multiple-byte multiply operations by itself.

Rotate is a great way to receive or transmit *serial data*. Letters like the ones you see on this page are sent in an eight- or seven-bit code called *ASCII*. These letters are handled by computers and microcontrollers in a byte format. When computers communicate with each other, or with printers, they frequently use serial transmission. Serial transmission is sending the information through a communications link bit by bit. You can see that this would involve the translation of the byte to sequential bits. To transmit, you would rotate left the byte with carry, and set the output pin high or low depending on the state of the carry bit. Do this eight times, and the serial transmission is complete. To receive, set the carry bit high or low depending on the state of the input pin, and rotate right the byte with carry. Repeating this eight times will complete the reception of the byte. True serial communications is more complicated because of the timing and extra bits required to meet standard protocol. The standard for PC communications is called RS-232. In one of the projects in this book, you will construct an operational RS-232 port.

CHAPTER 3
Bits and Bytes

Chapter 2 was primarily a generic discussion of digital logic. It can be applied to anything from TTL chips to the largest mainframe computers in use today. You will now see how this information is implemented in a microcontroller. As explained, all digital electronics is based on the bit. Eight bits are grouped together into a byte (for an 8-bit controller). Microcontrollers function by the manipulation and storage of these bytes. There are many different, highly-specialized internal parts of controllers which handle either manipulation, storage or control. Indeed, before the single-chip microcontroller, these subsections were separate components. Some microcontrollers today still use external ICs to extend their functionality.

You may be asking why such an in-depth knowledge of the internal workings of the microcontroller is needed. After all, most of us did not have to disassemble an engine to learn how to drive a car! The difference lies in the need to program the controller to make it work. All instructions deal with manipulating or moving bytes between internal modules, such as the registers or RAM. Obviously, without knowing what these modules are and how they work, you could never write a control program. Some controllers, thanks to the use of high-level languages like C, insulate us from the underlying hardware, but only to a certain degree. Knowledge of I/O ports and interrupts at the hardware level are still required to do any useful controlling.

Once a particular microcontroller is learned, this experience can be carried forward toward learning the next controller. However, controller families are different enough from each other that each family line requires the study of the idiosyncrasies of its architecture. Many elements are the same; consequently, the learning curve is less for each successive controller.

So, let's proceed and find out what makes a microcontroller tick.

Registers

Registers are special purpose memory locations. Memory is defined as the ability to retain a value over time (usually until power is turned off). Traditionally, all memory inside the processor is referred to as registers, but some locations have special functions, and are referred to as special purpose registers. In the parlance of 80C51, special function registers (SFRs) occupy

SYMBOL	NAME	ADDRESS
ACC	Accumulator	0E0H
B	B Register	0F0H
PSW	Program Status Word	0D0H
SP	Stack Pointer	81H
DPL	Data Pointer Low Byte	82H
DPH	Data Pointer High Byte	83H
P0	Port 0	80H
P1	Port 1	90H
P2	Port 2	0A0H
P3	Port 3	0B0H
IP	Interrupt Priority Control	0B8H
IE	Interrupt Enable Control	0A8H
TMOD	Timer/Counter Mode Control	89H
TCON	Timer/Counter Control	88H
T2CON	Timer/Counter 2 Control	0C8H
THO	Timer/Counter 0 High Byte	8CH
TLO	Timer/Counter 0 Low Byte	8AH
TH1	Timer/Counter 1 High Byte	8DH
TL1	Timer/Counter 1 Low Byte	8BH
TH2	Timer/Counter 2 High Byte	0CDH
TL2	Timer/Counter 2 Low Byte	0CCH
RCAP2H	T/C 2 Capture Reg. High Byte	0CBH
RCAP2L	T/C 2 Capture Reg Low Byte	0CAH
SCON	Serial Control	98H
SBUF	Serial Data Buffer	99H
PCON	Power Control	87H
IOCON	IO Control	F8H

Table 3-1. *A table of the special function registers (SFR) in the 80C51.*

memory locations from 128 to 255. The number of the memory location is its address, similar to the address of a house. Not all of these addresses have implemented registers. In the 80C51, only certain instructions can affect the SFRs; some of these instructions cannot affect general purpose memory. A good example of an 80C51 SFR is the accumulator. Mathematical operations must use the accumulator as one of the operands and location of the result. In the PIC® line, all memory locations can be used as the accumulator. *Table 3-1* shows the SFRs in the 80C51. Later, you will see in detail what some of these SFRs do.

You have seen how bytes are addressed. Sometimes bits can be addressed as well. This is especially common in registers. One example which you have already seen is the carry. A bit that is addressable is often called a *flag*. In the 80C51, the carry flag is the most significant bit of the program status word (PSW, byte address hex D0). In programming, these flags are addressed by special symbols. The carry flag is simply C. Proper bit addressing in most assemble languages is the appending of a period followed by the bit number. Thus, the carry flag can also be addressed as D0.7 or PSW.7. *Table 3-2* shows one SFR, the PSW, and how it is divided into a number of flags. Not all register or memory locations are bit-addressable. An exception to this is the PIC® line, where all bytes have addressable bits.

SYMBOL	NAME	ADDRESS
CY	Carry Flag	PSW.7
AC	Auxiliary Carry Flag	PSW.6
F0	Flag 0 available to the user for general purpose	PSW.5
RS1	Register Bank selector bit 1	PSW.4
RS0	Register Bank selector bit 0	PSW.3
OV	Overflow Flag	PSW.2
F1	Flag F1 available to the user for general purpose	PSW.1
P	Parity Flag. Set/cleared by hardware each instruction cycle to indicate an odd/even number of '1' bits in the accumulator	PSW.0

Table 3-2. One SFR can hold up to eight flags. Each flag has important functions in the microcontroller.

In the PIC® 16C54 - 16C57, the first eight bytes of memory are special purpose registers. Register 06 represents the I/O pints on port B. Each bit in port B can be configured as an input or an output (see the section on I/O ports). Let's say all of port B is output. By moving the byte AA (10101010) to register 06, you have set alternating pins of port B to +5 volts or 0 volts. Therefore, a line *setb 6.7* in code will set the most significant bit of port B high, without changing the other bits in the port.

RAM

RAM stands for *random access memory*. RAM can be written to and read from. Almost all registers are RAM. RAM is used for general program or data storage; special purpose registers are not referred to as RAM. In PICs®, after memory location 07, there can be up to 72 bytes of RAM (or general purpose registers). The 80C51 has 128 bytes of RAM. Please bear in mind that there are many versions of microcontrollers in a family line. There are versions of the PIC® and 80C51 which have more RAM (and other functions) than those mentioned above.

RAM can be static or dynamic. Both types will retain memory only while power is present. Dynamic memory requires a periodic refresh to each memory location or memory is lost. This refresh is often invisible to the user, or is provided by a dynamic memory controller chip. With microcontrollers, static RAM is common.

External RAM can be provided for most controllers. Some architectures were designed with this in mind, with the ability to change I/O ports into standard address and data lines. (See Chapter 13 for an example of address and data lines.) The 80C51 does this very well and can accommodate an extra 65,000 bytes of RAM. Obviously, by using the I/O ports for address and data lines, you are transforming the microcontroller into a computer processor! This is not the primary purpose of a microcontroller, and we shall stay away from such multi-chip applications for now. One of the later chapters will show the possibilities behind adding RAM, ROM, serial chips, video controller and I/O ports to a microcontroller bus.

The PIC® line was not designed with the concept of adding RAM. Even so, there are two ways of doing this. A software routine can simulate the proper address and data lines for the connection to a standard static RAM. The second method is RAM that can be interfaced using a three-line serial interface or the I²C bus. Once again, a software interface is necessary, but the advantage is that you are not typing up I/O lines for communications with the RAM.

Okay, now you know what RAM is, and where it is, but why do you need RAM? If you are going to do something in a loop, say 50 times, somewhere you have to keep track of the iteration number. You could put 50 into a RAM byte and decrement it for each loop. To do averaging requires the temporary storage of numbers before the averaging takes place. Complex mathematical routines require many bytes of memory. Serial communications use input buffers and output buffers. Buffers are sequential bytes in RAM of various lengths. If the processor is busy, interrupt-driven serial communications will store incoming bytes in the buffer until they can be properly handled. Basically, all programs use RAM and you shall see many examples of RAM usage later in the book.

ROM

ROM is read-only memory, and it is used to contain the program code and data lookup tables. ROM retains its content even when power is removed. ROMs are created in semiconductor factories by the thousands from a *mask*. A mask is the information of which byte will hold what value. When done in high quantity, the per-piece price is very low. Of course, the disadvantage is the inability to change the data in the ROM or to do small runs. For all practical purposes, this eliminates the use of ROM for small jobs, for R&D (research and development), or for the hobbyist. Instead, you will use EPROM, which is covered in the next section.

This book has mentioned the 80C31 and the 80C51 several times, often interchanging them. They are the same part except for the ROM. The 80C51 has internal room for 4K (K represents 1000) bytes of ROM. During R&D, you should use the EPROM version of the 80C51 (87C51). After the code is perfected, a mask is created and the permanent 80C51 is produced. Again, in high quantities, this is the most cost-effective way to work. In smaller production runs, the next best solution is OTP (one-time programmable). This is similar to the EPROM version of the 80C51 except, as the name implies, it can be programmed only once. The last option is to use the ROMless 80C31 and match it with external ROM. When external ROM is used, design changes are necessary to accommodate external address and data lines.

An external ROM is also used when the internal ROM capacity is not enough for the application. The 80C51 has internal 4K bytes of ROM, but can support up to 64K of external ROM. By tying a pin low on the 80C51, you can force it to fetch all program instructions from the external ROM (in essence, making it a 80C31). For higher-end industrial jobs, you can often see ROMs next to the processor or microcontroller. Typically, they are in a 24- or 28-pin DIP package, socketed on the board. This gives manufacturers the much-appreciated ability to update the product by just replacing the ROM.

EPROM

Closely related to ROM is EPROM, erasable programmable read-only memory. EPROM can be substituted anywhere ROM is used. The main advantage is that you can erase the program or data and use it again for the new version of the program or data. During the development of a microcontroller project, it is necessary to rewrite EPROM several times to fully debug a project. EPROM comes in many different sizes, from a few hundred to hundreds of thousands of bytes.

EPROMs are erased by ultraviolet (UV) light, which is why chips with EPROM in them have a clear, round window in the middle. Ultraviolet light is not a primary element of household light or sunlight. (Leaving an EPROM under an ordinary light bulb would erase its memory in about a year, or under sunlight, in approximately a month.) A special UV eraser can be purchased from electronic supply houses for about forty dollars, and is an indispensable tool for anyone who works with microcontrollers or EPROMs. A less expensive alternative is a UV bulb of the type used for house plants, or a black light, but be careful not to stare at the UV source as it is harmful to your eyes.

Microcontrollers come in versions called *one-time programmable* (OTP). OTPs cost much less money than EPROMs that can be erased more than once. An example would be the PIC® 16C57; with EPROM, $22.35 versus the OTP cost of $11.25 (with a high-speed oscillator option). It makes sense to first use EPROM then switch to OTP for limited production runs. The reason for the cost difference is surprising. Even though the internal chip is the same between EPROM and OTP, what makes the difference is the package! With OTP, cheap plastic carriers are used, whereas the EPROM version has a ceramic and clear-window package. Without the window, even though the OTP's insides are erasable, the UV can't get in to erase the chip!

Closely related to EPROM is *electrically-erasable programmable read-only memory* (EEPROM). As the name suggests, this ROM is programmable and erasable by electrical means. While similar to RAM because you can read and write to it, it also has the ROM's property of retaining data stored in memory when power is off. This places it in a category called NV-RAM (nonvolatile RAM). It cannot be used as a direct replacement for RAM because the write cycle is much slower, and there is a limited number of times a memory location can be erased and rewritten. Several microcontrollers have EEPROM in them, and for those that don't, it is very easy to connect one to a controller.

EEPROM is tremendously useful in control applications for storing user-selectable options, data accumulation, and programmable set points. You could use EEPROM in almost all control projects: the following are just a few examples. Serial communications have different baud rates, parity, stop bits, and lengths which can be user-selectable. A terminal could have the ability to send any character or sequence of characters programmed by the user (stored in EEPROM). An alarm system might have programmable times for when a certain event goes off. Motion control requires the counting of distance and comparing it to a programmed preset distance.

When in doubt, include an EEPROM in your design. The interface requires only three I/O lines and the cost is as little as fifty cents in quantity.

I/O Ports

Input/output (I/O) ports are the primary method for interfacing the intelligence of a microcontroller with the real world. Each port is typically eight pins, though this is not always true. Most PIC® controllers have one 4-pin port. Each I/O port has an address and appears as a memory location when accessed. Any I/O pin can be configured for input or output. This is accomplished by writing a configuration byte to a special register. In the PIC® family, the command to set I/O pins is called TRIS. A high bit represents input, a low is output. Thus, the command *TRIS 7,11110000b* sets port C (7), with the high nibble all-input and the low nibble all-output.

When configured as input, the I/O pin is in a high impedance mode. The device driving the pin must provide sink and source capability. What this means is that if nothing was attached to the I/O pin, it would float. The voltage could be anywhere between 0 volts and 5 volts, somewhat randomly depending on certain conditions. Next, hook up the output of a device to this pin so you can read it. One type of output is called an "open collector." It can sink current to ground, but does not provide a source of current to drive the pin high. *Figure 3-1* shows an open collector output. The solution to make this work is not difficult; add a pull-up resistor to provide source current. Pull-up resistors are also quite commonly used to read switches. When the switch is open the input pin is 5 volts, and when the switch is closed, the pin is ground. (See *Figure 3-1.*) Without the pull-up, when the switch is open, the pin would float, giving you erroneous readings.

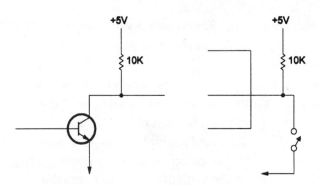

Figure 3-1. *An open collector transistor and a pull-up resistor to provide source current.*

Let's look at the conditions of an I/O pin that is configured as an output. Almost all micro-controller I/O pins configured as outputs can provide both sink and source capability. *Figure 3-2* shows the internal workings of an output pin. By turning on the P MOSFET, you provide 5 volts on the output, while turning on the N MOSFET provides ground. (During input, both of these are turned off.) This kind of internal detail is invisible to you; you simply send a high or low to the pin. Do not attach the output pin to ground or a low resistance to ground, or damage can occur to the microcontroller. The specification for each controller tells you what kind of load the I/O pin can drive. In the case of the PIC® 16C57, any one pin can sink 25 mA or source 20 mA. Total current per port cannot exceed 50 mA or 40 mA, respectively. The total current is important to watch as you can easily rise above 50 mA with three or more pins outputting simultaneously. Unless you are sure no more than one pin at a time will be active, always design the load for each pin to be 1/8th of the port maximum.

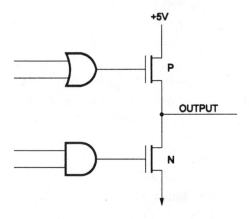

Figure 3-2. *The output stage of an average microcontroller. MOSFETs provide both source and sink current.*

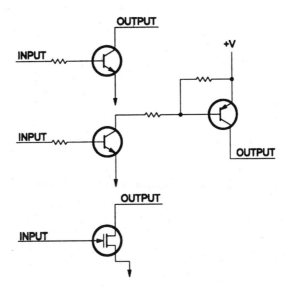

Figure 3-3. Trying to drive too much current with the microcontroller output will destroy the chip. Here are a few methods for increasing the current handling capacity of a microcontroller.

What if you need to drive relays, high current LEDs or other devices that exceed the maximum load per pin? Buffer transistors or MOSFETs are good, inexpensive solutions. *Figure 3-3* shows several different buffers for high current output.

Interrupts

An interrupt is exactly what the name implies; it interrupts the current executing program. Some microcontrollers have one or more pins to allow for an incoming interrupt source. The low end of the PIC® family has no interrupt capacity at all. Be aware that there are software interrupts as well, but this section discusses only hardware interrupts.

The interrupt can be programmed to react to a high level, a low level, or to the edge of a changing signal. When the signal on the interrupt pin matches the configuration, an interrupt is issued. Microcontrollers act differently at this point. Some simply store the executing address then jump to the interrupt routine. Others store the current address plus several important registers before jumping. The reason for storing registers is to save data that is being used for the current program. You never know when an interrupt is going to come in—

one might occur in the middle of a mathematical routine using a temporary value in the accumulator. The interrupt routine has a good chance of changing this value. By saving the register, the microcontroller returns the value upon leaving the interrupt routine. Controllers that do not save registers must have carefully-written interrupt code so as to not to destroy any needed data. Interrupts are usually confusing to the beginner. One of the projects in this book shows interrupts in action.

Some microcontrollers have several interrupt pins. This allows for interrupts from a number of sources. The pins have a hierarchy, and some are more important than others. If you are executing an interrupt from a high level pin, and an interrupt comes in from a lower level, the controller will ignore it. If a low level interrupt is running and a high level interrupt comes in, then the controller will interrupt the interrupt. You can also turn off all interrupts in case the processor is executing an important segment of code. The only exception to this is the *non-maskable interrupt* (NMI), which will always execute.

What if you have a number of devices that will generate interrupts but only through one interrupt input pin? If all the devices have open collector outputs and go low on interrupt, then you can tie the outputs together to the input pin with a single pull-up resistor. With any other circumstances, feed the interrupts through inverters if it's required to achieve a constant

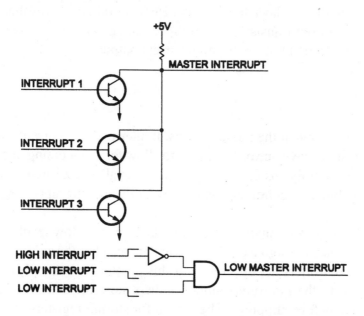

Figure 3-4. ORing interrupts create a master interrupt to the microcontroller.

trigger level, then AND or OR the lines into the input pin. (See *Figure 3-4* for examples.) Now, any one of the devices will trigger the interrupt. You might ask, how will the controller know which device initiated the interrupt? The interrupt routine must poll each device until it finds which one set off the interrupt.

Some microcontrollers have no interrupts at all. This may seem like a disadvantage, and it is for time-critical applications. Polling to check the status can make up for interrupts in a large number of projects. Several of the projects in this book use polling, and you will see how efficient it can be for the replacement of interrupts.

CHAPTER 4
Analog-to-Digital
Digital-to-Analog

Working with the real world requires more than highs and lows. You have to deal with varying light levels, volume levels, frequencies, amplification levels, and more. This realm of variable voltage is called analog. How can the microcontroller, with only two possible voltage levels, handle the hundreds or even thousands of levels required in analog applications? It can incorporate an analog-to-digital (A/D) interface as a internal module, or a separate IC. The voltage level is sampled and translated to a number, which is then stored in digital format. The output module is a digital-to-analog (D/A) interface that can also be internal to the controller, or separate. The digital number (stored as a byte or a word) is translated by the D/A to a single voltage level.

We shall also look at sensors that can feed the A/D or receive voltage from the D/A. Without sensors, your devices would not react to any real time events—a primary purpose of embedded control applications.

A/D

The exact inner workings of the A/D are beyond the scope of this book. Suffice it to say that a voltage between two points (minimum and maximum), when input to the A/D, will produce a number in binary. Depending on the type of A/D, this number could be 8, 12, 16, or other bits long. A/Ds that produce 8-bit results are very common. The greater the number of bits, the more resolution in the transformation. To see this, let's use the example of a voltage level between 0 and 5 volts. With an 8-bit A/D, there are only 256 possible divisions for the 5-volt span. By dividing the 5 volts by 256, you can see that the maximum resolution is 19.5 mV. If this had been a 12-bit A/D, you would have had 4096 divisions and a maximum resolution of 1.2 mV.

Figure 4-1 illustrates the results of an 8-bit A/D transformation. The sinusoidal waveform falls between 0 and 5 volts. Five different samples are taken. A/Ds can be configured to sample at a periodic interval; but more often than not, the sampling is done under microcon-

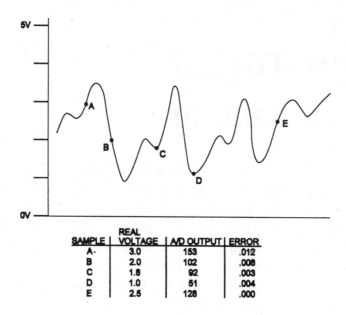

SAMPLE	REAL VOLTAGE	A/D OUTPUT	ERROR
A.	3.0	153	.012
B	2.0	102	.006
C	1.8	92	.003
D	1.0	51	.004
E	2.5	128	.000

Figure 4-1. *Values from an A/D conversion at the points indicated on the waveform.*

troller control, and occurs irregularly. An external A/D provides the samples to the controller by either a parallel bus (wasteful of I/O lines, but quick) or a serial interface (uses far fewer I/O lines than the parallel bus, but is ten times slower). The microcontroller then processes the A/D samples or stores them for later use. You can also see in *Figure 4-1* that the inherent resolution error is small enough to be ignored for most applications.

The speed of A/D conversion is important for two reasons. Some A/Ds do not have a built-in "sample-and-hold" circuit. Sample-and-hold (sample/hold) freezes the input as the conversion process is taking place. Without this capability, the input voltage would change while the conversion was running! Normally, the conversion speed is fast enough that this does not matter. However, let's consider the case of a common A/D, the ADC0831. The maximum conversion speed is 32 µs. For an accurate sampling, the input voltage must not change more than 19.5 mV in 32 µs. If it did, then the signal would change levels during the conversion process. 19.5 mV in 32 µs translates to 61 hertz for a 5-volt peak-to-peak signal. This is slow enough that you would not want to sample a frequency (such as music from a microphone), but rather a fairly constant voltage level (dc). Other examples of A/Ds are the ADS7819 (12-bit, sample/hold, 1 µs conversion rate), ADS7808 (12-bit, 10 µs conversion rate) and ADS7805 (16-bit, sample/hold, 10 µs conversion rate).

The sampling rate mandates the greatest possible *frequency* that can be sampled. If a conversion process takes 32 µs, then you cannot sample faster than 31,250 times per second. In reality it is slower than this because of overhead used for communicating with the A/D, plus microcontroller instruction time. If the objective is to analyze the frequency of the input signal, care must be given to sample the input quickly enough. *Figure 4-2* shows what happens when the sampling frequency is less than the half the frequency of the input. The result is called a beat frequency—a perfect sine wave at a lower frequency, and very misleading. To do frequency sampling, the maximum frequency that can be correctly found is referred to as the *Nyquist frequency*, or the sampling rate as divided by 2. This is an important limitation when using *fast-fourier transformation* (FFT). FFT is used to analyze a waveform and deliver the fundamental frequencies comprising it. Is this too much for a simple microcontroller? No. Right now there are microcontrollers using A/D and FFT to process telephone touch-tones and status signals.

D/A

D/A is the exact opposite of A/D: a byte is converted to an analog voltage level. While it is rare to see the D/A module incorporated into the microcontroller, external D/As can be interfaced to the controller by either serial or parallel communications. A typical D/A output swings between ground and 5 volts (or the voltage level of the microcontroller). Some D/As have a pin that enables the analog level to be much higher than the logic level. The resolution depends on the voltage span divided by the number of bits (8, 12, 16, etc.). Thus, an 8-bit, 5-volt D/A will have maximum resolution of 19 mV.

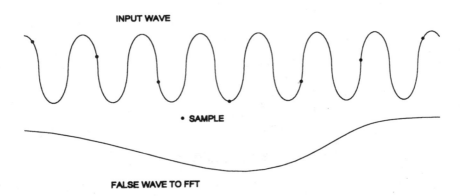

INPUT WAVE

• SAMPLE

FALSE WAVE TO FFT

Figure 4-2. In performing a FFT, do not exceed the Nyquist frequency or very misleading results can occur.

In the preceding example, each output level of a D/A is a discrete rise in voltage (19 mV). Instead of being a smooth and continuous rise, the voltage level resembles the steps in a stairwell. When the output is viewed on a oscilloscope, the sine wave from a D/A is very ragged. To cure this, a filter is often included at the output of a D/A. The filtering can be complicated or be as simple as an RC filter, or even a simple capacitor. The nature of the filter depends on the desired waveform from the D/A.

A typical *digital-to-analog converter* (DAC) is the Maxim MAX522. It comes in an 8-pin package and features two independent analog outputs. The MAX522 has 8-bit resolution and can operate off a single voltage supply. Its three-wire serial interface makes it a simple job to interface with any microcontroller. You will see one example of a three-wire interface later, both the hardware and software.

The standard D/A chip provides for input buffering and output isolation—nice features in most applications. However, if you have extra I/O lines and low cost is a factor, then consider this interesting alternative: an R-2R ladder network will work as a passable D/A. *Figure 4-3* shows a 4-bit R-2R ladder. The name is derived from the use of only two resistors, R and 2R. (Example, 5K and 10K.) Note that the impedance along the ladder is the same for each input bit. This can be readily seen by the following analysis: if bit PC0 is ground, then resistors D and E are in parallel reducing to a value of R (2R and 2R in parallel is equal to R). R is added to F, producing 2R, again in parallel with resistor C once more producing R. This is continued along the length of the ladder for any number of bits. The impedance always remains the same. As an example of the output, let's look at a microcontroller with the

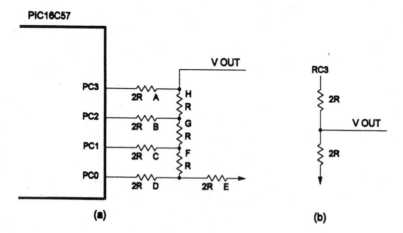

Figure 4-3. *A crude (but cheap) D/A can be created with the resister in the R-2R configuration.*

number 8 on the R-2R interface I/O lines. In binary, an 8 is 1000. This sets 5 volts on PC3 and 0 volts on all the other pins. From the previous analysis, you know the ladder collapses to the equivalent of *Figure 4-3b*. This produces 2.5 volts as the output, which is correct for the nibble. While the R-2R ladder is not linear, it will suffice as a cheap and easy D/A. One application uses it for an acceptable human voice with an appropriate amplifier for volume.

Sensors

Sensors are to microcontrollers what your eyes and ears are to you. Microcontroller sensors can detect and quantify movement, speed, light, pressure, moisture, temperature, sound, material and distance. All sensors fall into two basic categories: those that output a digital logic level, and those that output an analog result. The most fundamental sensor is the switch. With a switch, you can set limits of travel, detect doors being opened or closed, read human input and device positions, and countless other uses. A switch is simple to implement and is fully microcontroller compatible. Hook one side to the controller with a pull-up resister on the line, the other side to ground. A high on the input indicates that the switch is open while

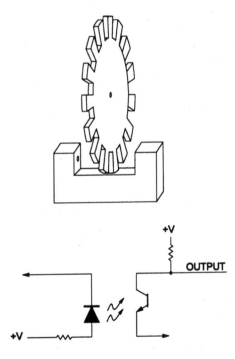

Figure 4-4. *A slot sensor can be used to read the notches on a wheel.*

a low means that the switch is closed. Closely related to the mechanical switch is the reed relay. This 1/2-inch long tubelike device contains a switch that is activated by the presence of a magnet. Reed relays are popular for use as door sensors in an alarm system. On the door itself is a fix magnet, while on the door jam is the reed relay with wires going to the central control box. The reed relay can be replaced with a *hall-effect transistor*, a solid-state (like a transistor) equivalent device. (Beware of voltage and current limitations.)

Right behind switches are optical emitters and photo detectors. A popular form is the interrupter module, sometimes called a slot sensor. *Figure 4-4* shows a basic configuration of a slot sensor in action. As with all emitter-detector pairs, the emitter generates light that is received by the detector. As shown in *Figure 4-4*, when light passes through the slot, the detector is on, connecting ground to the output; when the light path is blocked, the line is pulled high. This high/low is directly readable by a controller, though care must be given to ensure a clean digital signal. Photo detectors can be purchased that include digital buffering on the output for this purpose—they are called *logic detectors*. There are many other uses for an interrupter module. A metal flange on a device in linear motion can break the light path to indicate the end of travel. It could also be a paper or material-present sensor (paper passing through the slot sensor).

Among the many analog sensors are those that sense temperature and light. A light sensor is a photo detector (sensitive to the visible light spectrum) with appropriate amplification. The output is read by an A/D, and the relative intensity is then output as a byte value to the microcontroller. This has applications in automatic light controls and burglar alarms. A temperature sensor is similar to a photo detector except that it reacts, predictably enough, to temperature. Again, amplification and an A/D are required to interface to a microcontroller. One of the more complicated projects in this book is an indoor/outdoor video thermometer. In this project, you will use a compact temperature IC. The chip will have the temperature sensor, amplifier, A/D, and three-wire interface, all in one 8-pin chip!

We've only discussed some of the more popular sensors in this section while many others have been skipped. They are all useful, and an entire book could be written just about sensors. In time, your increased experience and the need to use microcontrollers in a wide variety of applications will expand your horizons to the numerous possibilities.

CHAPTER 5
The PIC® Family
of Microcontrollers

By now you should have the basics well understood and are ready for a more in-depth look at the microcontroller. There are so many controllers, most excellent in their own way, that this book would need to be expanded to several volumes to cover each one. The similarities between different microcontroller families enable us to do a concentrated study on one and apply that knowledge to others. But which one? As mentioned earlier, we will be focusing on the Microchip family of PIC® microcontrollers. This does not mean that PICs® are any better than other controllers for a certain application; rather, in the difficult task of choosing the best one for this book, emphasis was given to the learning process. PICs® will enable you to learn about microcontrollers; but in any serious application, you should consider the vast array of the different controllers on the market today.

PICs® are easy to acquire, inexpensive, have low cost development tools, and are powerful. Popular nationwide distributors such as Digi-Key and Mouser carry a full line of PICs®. For large purchases, several specialized distributors are available that sell PICs® at rock-bottom prices. Even in single quantities, the previously mentioned vendors sell OTP PICs® for just a few dollars. As with all microcontrollers, you can't do much without development tools. Compilers, programmers, emulators (all of which will be covered shortly), are abundant and very inexpensive. It is even possible to build for yourself the most basic tool: a programmer. You can get all of these advantages without shortchanging yourself in performance. With an MIPS of 5, there is almost nothing you can't do with the PIC®.

Microchip

Microchip Technology, Incorporated, is headquartered in Chandler, Arizona (near Phoenix). At this facility, they have their executive offices, R & D, and wafer fabrication facilities. In 1993, they purchased a second wafer fabrication building in Tempe, Arizona. Microchip certainly needed this added capacity as their microcontroller line alone has sold over 100 million units to date. They are expanding at a rate of 25% per year. Microchip is in the worldwide market with sales offices in cities throughout the Western Hemisphere, Europe, and the Pacific Rim.

Along with microcontrollers, Microchip produces other integrated circuits. They have focused on chips related to controllers and microprocessors. One family of chips they manufacture is serial EEPROM, ranging from 256 to 64K bits. They also produce parallel EEPROM (not really useful for microcontrollers). Microchip makes EPROM ICs, compatible with standard EPROMs, ranging from 8K to 64K bytes. They do an occasional odd logic IC, such as the AY0438 (32-segment CMOS LCD driver). Using their PIC® controller as the core, Microchip has produced a line of ASICs. These ICs are used for battery management and mouse controllers.

Without a doubt, Microchip's main line is microcontrollers. They have a diverse set of products, all based on the RISC architecture. Their controllers differ mainly in the number of I/O lines, oscillator types, power consumption, memory, and internal modular configuration. *Table 5-1* lists Microchip's line of microcontrollers. You can see the variety of microcontrollers that are available in the PIC® family. Bear in mind that the PIC® family is changing even as you read this; it will undoubtedly contain even more versions, so contact Microchip for the latest information. We will now look at the hardware of a few PICs® in detail.

16C54

The PIC® 16C54 has a top speed of 20 MHz (5 MIPS), 12-bit wide instructions, an 8-bit data path, 512 bytes of program memory, 32 bytes of RAM (8 special purpose registers), a 2-level hardware stack, and 33 instructions. Peripheral features include 12 I/O pins, a real-time clock/counter (RTCC), a watchdog timer, a security program fuse, and four types of oscillator circuits. Similar to the rest of the PIC® family, the 16C54 uses CMOS technology, is fully static, has a wide operating voltage range of 3.0V to 5.5V, and a low power consumption rate of less than 2 mA at 4 MHz. *Figure 5-1* shows the pinout of the 16C54.

RA0 to RA3 are I/O pins of port A, and RB0 to RB7 are I/O pins of port B. Each pin can be defined as input or output. The RTCC real-time clock/counter pin is for the input of a digital signal. This signal is counted directly or divided by a programmable prescaler, all under user-configurable control. The MCLR master clear is active on a low, and will reset the microcontroller. This pin is normally tied to 5 volts. Under unusual circumstances (power comes up slowly, very low frequency oscillator), an RC circuit is attached to this pin. OSC1 and OSC2 pins are used for the generation of a clock signal. If you are using a crystal oscillator, the two

Chip	Speed (MHz)	EPROM bytes	ROM bytes	EEPROM bytes	RAM bytes	I/O	Serial Ports	Parallel Port	A/D	Interrupts Sources	PWM MOD	Instructions	Packages
PIC16C54	20	512	--	--	32	12	--	--	--	--	--	33	18 DIP, 18 SOIC 20 SSOP
PIC16R54	20	--	512	--	32	12	--	--	--	--	--	33	18 DIP, 18 SOIC 20 SSOP
PIC16C55	20	1K	--	--	32	20	--	--	--	--	--	33	28 DIP, 28 SOIC 28 SSOP
PIC16C56	20	1K	--	--	--	12	--	--	--	--	--	33	18 DIP, 18 SOIC 20 SSOP
PIC16C57	20	2K	--	--	80	20	--	--	--	--	--	33	28 DIP, 28 SOIC 28 SSOP
PIC16CR57A	20	--	2K	--	80	20	--	--	--	--	--	33	28 DIP, 28 SOIC 28 SSOP
PIC16C58A	20	2K	--	--	80	12	--	--	--	--	--	33	18 DIP, 18 SOIC 20 SSOP
PIC16C64	20	2K	--	--	128	33	Yes	Yes	--	8	1	35	40 DIP, 44 PLCC 44 QFP
PIC16C71	16	1K	--	--	36	13	--	--	4 ch	4	--	35	18 DIP, 18 SOIC 44 QFP
PIC16C74	20	4K	--	--	192	33	Yes	Yes	8 ch	12	2	35	40 DIP,44 PLCC 44 QFP
PIC16C84	10	--	--	1K	36	13	--	--	--	4	--	35	18 DIP, 18 SOIC 20 SSOP
PIC17C42	25	2K	--	--	128	33	Yes	--	--	11	--	55	40 DIP,44 PLCC 44 QFP

Table 5-1. A partial listing of Microchips's family of microcontrollers.

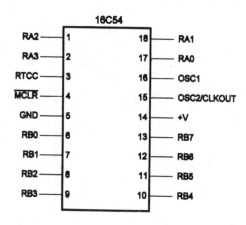

Figure 5-1. *The pinout of the "low end" of the Microchip PIC® family, the PIC® 16C54.*

leads are tied to OSC1 and OSC2. With the use of an RC timer, connect the common end of the RC to OSC1. OSC2 is then an output producing a square wave one quarter of the input on OSC1. *Vdd* is power (typically +5 Vdc) and *Vss* is ground.

Figure 5-2 shows a block diagram of the internal workings of the PIC® 16C54. The EPROM contains 512 bytes of memory that are used for program storage. The EPROM addressing is done by the PC program counter register. Directly attached to the PC is the stack. This is a two-layer hardware stack. The stack is used to keep track of where the program is after a subroutine call. This enables you to return to the calling location after finishing the routine. In this processor, the only way to access the stack is by doing a CALL or RETURN command. Other microcontrollers have directed control of the stack with PUSH (put a number on the stack) and POP (retrieve a number from the stack). Note that the PIC® is a *two-level* stack. Do not attempt to execute more than two CALLs without a RETURN, or else the program will malfunction. The PC is incremented sequentially with the exception of CALL, BRANCH and JUMP commands. The instruction from EPROM, addressed by the PC, is moved into the instruction register and instruction decoder. The instruction is then executed on the rest of the hardware.

The PIC® 16C54 has 32 bytes of RAM; seven are special purpose registers and the other 25 are general purpose registers. Register 0 is reserved for indirect addressing, which will be covered in the next chapter on PIC® firmware. The next register is the RTCC counter. The PC is register 2, and unlike the other registers, it is 11 bits long. Register 3 is the status register. The status register contains various flags for microcontroller operation. These flags

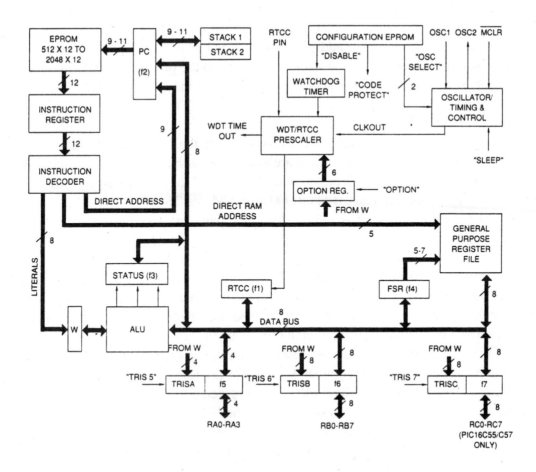

Figure 5-2*. A block diagram of the internal workings of the PIC® 16C54.* *
*Reprinted with permission of the copyright owner, Microchip Technology Incorporated © 1996.
All rights reserved.*

are *carry* (bit 0), *digit carry* (bit 1), *zero* (bit 2), *power down* (bit 3), and *time out* (bit 4). Register 4 is the FSR, file-select register. FSR is used in indirect addressing and memory bank selection (but not used in 16C54). It is always a good idea to set FSR to 0 at the start of the PIC® program. Registers 5 and 6 are I/O ports A and B. The rest of the registers can be used for general data storage while the PIC® is running. The memory is cleared when power is removed.

One of the most confusing aspects of the PIC® is the limitation on subroutine calls. The program memory of the 16C54 is 512 bytes (one page of memory). The instruction that does a call allows only an 8-bit address, and thus can address only 256 bytes. Simply put, only calls to the first 256 bytes of a page can be made. A call can be made from any location; this is because the stack which retains the address of the calling line is 11 bits long. The return from the subroutine uses the address in the stack and returns to the proper location. Some PICs® have four pages of program memory, but you can only execute calls to the first 256 bytes of each page.

Both I/O ports have special registers associated with them, the TRISA and TRISB. They are not in the normal memory of the PIC® and are accessible with only one command. TRISA and TRISB respectively sets which pin on port A and port B to input or output. A high (1) is input and a low (0) is output. Thus, if 11110000 is moved into TRISB, then the high nibble of port B is all input and the low nibble of port B is all output. An easy way to remember this is the shape of 1 and 0; **1** is **I**nput and **0** is **O**utput.

We have discussed the functionality of the RTCC. One of the options is to add a prescaler to the RTCC. In *Figure 5-2*, you can see that the prescaler is shared between the RTCC and the WDT (watchdog timer). The prescaler can be used only for one function at a time. The WDT is used as a method of assuring that the firmware program is running correctly. Basically, the WDT is a counter which, if not reset in a certain amount of time, resets the micro-controller. One of the controller commands is CLRWDT (clear watchdog timer). This instruction must be issued periodically if the WDT is turned on. If the controller gets stuck in a faulty loop, or malfunctions due to static shock, the WDT will reset the microcontroller. The WDT is very useful and should be used for any serious application. The nominal time-out period is 18ms without a prescaler, and a maximum of 2.5 seconds with a prescaler.

The PIC® can run with four different types of oscillator circuits. With EPROM PICs®, this choice is selectable, while OTP must be purchased with the correct oscillator in mind. An external source can provide the clock as input to the OSC1 pin. (It must have correct voltage level and frequency.) Either a crystal or a ceramic resonator can be connected to the OSC1 and OSC2 pins. When using this type of oscillator, small capacitors (15 - 300 pf) must be connected from OSC1 and OSC2 to ground. The PIC® clock speed is the crystal frequency divided by 4. The most common (and inexpensive) clock generator is the RC oscillator. Connect a resistor between the +5 Vdc and OSC1 pin, plus a capacitor between the OSC1 pin and ground. Keep the resistor value between 2.2 Kohm and 1 Mohm. Use a capacitor above 20 pF. *Table 5-2* shows the oscillator frequencies for different RC combinations. The PIC® clock speed is the RC frequency divided by 4. This clock can be seen on the pin OSC2.

Capacitor	Resistor	Frequency	Tolerance
20pf	3.3K	4.71 MHz	± 28%
	5K	3.31 MHz	± 25%
	10K	1.91 MHz	± 24%
	100K	207.76 KHz	± 39%
100pf	3.3K	1.65 MHz	± 18%
	5K	1.23 MHz	± 21%
	10K	711.54 KHz	± 18%
	100K	75.62 KHz	± 28%
300pf	3.3K	672.78 KHz	± 14%
	5K	489.49 KHz	± 13%
	10K	275.73 KHz	± 13%
	100K	28.12 KHz	± 23%

Table 5-2. Different RC combinations produce different clock frequencies. *

*Reprinted with permission of Microchip Technology Incorporated © 1996. All rights reserved.

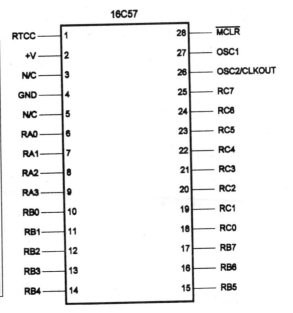

Figure 5-3. The pinout of the PIC® 16C57. Note the larger size and extra I/O lines compared to the 16C54.

16C57

The 16C57 is very similar to the 16C54. The following is a description of those items that differ from the 16C54. The size of the chip has grown from 18 to 28 pins. *Figure 5-3* illustrates the pinout of the 16C57. There are three primary additions: RC0 to RC7 is a third I/O port (C); the 16C57 has expanded its general purpose registers to 72; the program memory is 2K bytes.

ADCON1 Bit 1 Bit 0		RA0, RA1	RA2	RA3	Vref
0	0	A/D	A/D	A/D	Vdd
0	1	A/D	A/D	ref input	RA3
1	0	A/D	I/O	I/O	Vdd
1	1	I/O	I/O	I/O	Vdd

Table 5-3. The register ADCON1 (address 88h) in the PIC® 16C71.

Register 7 is port C. (In the 16C54, this was a general purpose register.) TRISC supports the new I/O port. TRISC will set the I/O pins to input or output, depending on the configuration byte. High bit for input, low bit for output.

By adding 48 bytes in three banks of 16 bytes each, you now have 72 bytes of general purpose RAM. These banks overlap the last 16 bytes of the first 32 . Access to the entire memory is accomplished by a scheme call *bank swapping*. The first 16 registers are always available. To use the extra memory, you must first set the bank select bits in the FSR register. These are bits 5 and 6. A 00 would be the first bank, 01 the second bank, 10 the third bank, and 11 the last bank. As an example, let's read/write the last byte (79). First, swap the fourth bank in by writing 11XXXXX (X means a bit that can be either high or low) to the FSR. Then read/ write to register 31. Why 31, you might ask? Because this where byte 79 appears after the bank swap. This is admittedly tricky, and we shall endeavor to work with only the first 32 registers in the firmware.

The program memory is now 2K bytes. Since the instruction for jumps has only 9 bits (512 possible address locations), how can you move between different parts of the program? We previously discussed the status register in the 16C54, bits 0 through 4. This remains the same in the 16C57, but you add the use of bits 5 and 6 as page select bits. When a jump is executed, these bits are transferred to bits 10 and 11 of the program counter. It is up to you, the programmer, to keep track of these address bits. Consider the case where the program is executing sequentially from byte 511 (page 1) to 514 (page 2). The program now wants to jump to a segment in the middle of page 2. After doing the jump, you find yourself in the middle of page 1! Even though you were in page 2, the page select bits in the status register were still pointing to page 1. Executing the jump wrote those bits back to the PC (which *had* incremented to page 2) and sent you back to page 1. Keeping track of what page you are on, and if you are in the first 256 bytes of that page (for subroutines), is very important in PIC® programming. We shall see examples of how to handle this later in the book.

16C71

With the 16C71, you shrink the chip back to 18 pins but add some important functionality. For the first time, you are looking at a microcontroller with an internal A/D (four channels) that is interrupt capable (four types of interrupts). *Figure 5-4* is the pinout of the 16C71. The only way to add all of these features with so few pins is to share the use of the pins. As an example, each pin of I/O port A can be configured to be an A/D input or an I/O line. We shall look at this in detail shortly. The 16C71 has 1K bytes of program memory and 48 bytes of

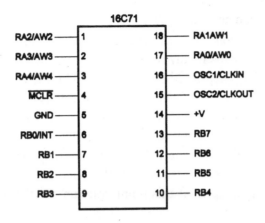

Figure 5-4. *Pinout of the PIC® 16C71.*
Several of the I/O pins can be configured as analog inputs.

RAM. A very nice feature of the 16C71 is the enlarged stack of 8 bytes. Implementing interrupts would be almost impossible without it. (Each interrupt uses one layer of the stack.)

There are four types of interrupts in the PIC® 16C71. Pin 6 can be configured as an interrupt from an external source. Inserting a magnetic stripe card into a reader with the interface tied to an external interrupt pin would be one example. When timer 0 overflows, it can be configured to be an interrupt. A good example of its use is when doing serial communications. Load the timer with the length between bits and process other information—when the interrupt hits, then the next bit is ready to go. The end of the conversion from the A/D module can trigger an interrupt. This allows the microcontroller to work on other jobs instead of waiting for the conversion to the end. An interrupt can also be issued when the bits on port B (bits 4, 5, 6, 7) change state. This means that any of them can change from a high to a low, or a low to a high.

Interrupts can be enabled or disabled individually or globally. The register that handles this is the INTCON register, address 0Bh. Bit 3 enables port B interrupts, bit 4 enables the pin 6 external interrupt, bit 5 enables the timer 0 interrupt, and bit 6 enables the A/D interrupt. Bit 7 is the global enable (GIE); it will enable or disable all interrupts regardless of the individual interrupt enable bit. All enable bits use 0 to disable and 1 to enable. Also in the INTCON register are flags to tell you which interrupt has gone off—bit 0 for port B, bit 1 for external, and bit 2 for timer overflow. A high in these bits means that its respective interrupt has gone off. The flag for A/D interrupt notification is bit 1 in a different register: the ADCON (register 8).

What is the sequence behind an interrupt? Let's go through one step by step. The microcontroller is processing data when the logic level changes on pin 6, the external interrupt pin. First, the GIE is set low to prevent any other interrupts. Then the current address (what the controller is currently doing) is pushed on the stack. The PC is then loaded with an address of 4, the interrupt service routine. Keep in mind that this is a firmware routine that you are writing. All interrupts go to the same address. It's a fair question to ask, how does the microcontroller know how to respond to the right interrupt? This is exactly what the interrupt notification flags are for. At the start of the interrupt routine, check each flag (see INTCON register) to see which one is set. The microcontroller can now perform the tasks necessary to handle the interrupt. At this point you can see the advantage behind the extra stack bytes. If you were in a second-level subroutine when the interrupt came, then pushing one more address onto a two-byte stack would be disastrous. Of course, the larger stack also gives you the ability to call other subroutines from the interrupt routine. Never go deeper than five calls from the interrupt routine. Now that the interrupt has been handled, it is time to return. Clear the external interrupt flag, set the GIE flag, and issue a return from the interrupt command. The interrupt is done.

A final word on interrupts: they are some of the trickiest areas to program. Learn the PIC® inside and out before attempting an interrupt routine. By its very nature, an interrupt can come at any moment. The controller may have just loaded an important piece of information into the work register and be about to use it when the interrupt comes. The service routine will undoubtedly change the value in the work register. Returning from the interrupt would then produce a serious problem. It is a good habit to save the work register in a general purpose register and restore it upon leaving the service routine. This applies to other important registers and illustrates the problems that interrupts can cause.

The 16C71 has four A/D inputs. They are multiplexed into one sample-and-hold and A/D converter. There are quite a few steps to completing your first A/D conversion. First, configure register ADCON1 (address 88h), selecting which inputs are A/D. (See *Table 5-3*.) Vref is the voltage range the A/D will work with.

Next, set up the register ADCON0 (address 08h) for clock, channel and A/D module. Bits 7 and 6 set the A/D clock as follows: 00 - chip clock/2, 01 - chip clock/8, 10-chip clock/32. 11 selects a separate internal RC clock of 4 μs nominal. The A/D clock must be over 2 μs. Select the channel by bits 4 and 3: 00 - AIN0, 01 - AIN1, 10 - AIN2, 11 - AIN3. Bit 0 of ADCON0 must be set high for the A/D to turn on. (Shutting off the A/D saves power.) At this point, you may turn on the A/D interrupt as described in the previous section. Start the conversion by setting bit 2 (GO/DONE) high. When the conversion is finished, this bit is set low by the hardware. The A/D result is in register 09h (ADRES).

Development Tools

To work with PICs® or other microcontrollers, you need tools. The most fundamental requirement is the programmer device. This can come as a self-sufficient stand-alone unit or a personal computer-controlled attachment. Next, and equally important, is the assembler. Then you have some nice development tools. Among your tools you should also have an emulator, a simulator, and a C language compiler. We will first look at offerings from the PIC® manufacturer, Microchip. There are many other companies providing similar products, typically referred to as "third party support."

All microcontrollers require firmware to run. Firmware must be *burned* (programmed) into the controller or external EPROM. This is exactly what a programmer module does. Microchip's high-end unit is the PRO MATE. It can operate as a stand-alone module or under control from a PC host system. In stand-alone mode, the unit can program controllers with firmware stored in internal memory, or it can read the program off a currently programmed PIC® and then program other chips. The PRO MATE can program a wide variety of PIC® controllers, including PIC® 16CXX and PIC® 17CXX.

A lower cost alternative is the programmer included in the PICSTART® package. This starter development system from Microchip contains a PC-controlled programmer, an assembler, a simulator, and two EPROM PICs®. The programmer is connected to a PC serial port. Software is included to run the programmer. The software has many functions, such as the ability to read an existing program out of a PIC®, store the program on disk, retrieve from disk, load object code, generate by assembler, verify PIC® content, directly edit program memory locations, and of course, burn the PIC®. A good programmer for beginners.

The firmware that runs a microcontroller consists of numeric codes. For example, 65 could mean to move the work register to register 12, or 73 could mean to load register 16 with I/O port B. The firmware could be written as 65, 73, 24, 35, etc., but the result would be very cryptic. This is where the assembler comes in. The assembler takes easy-to-understand mnemonics and translates them into the numbers required by the microcontroller. Thus, CLRW (clear work register) is translated to 40h and CLRF 9 (clear register 9) is translated to 69h. An assembler also takes care of housekeeping chores, such as constant references and address calculations. A constant is the same as a mathematical constant: a word or letters always representing the same number. Instead of writing CLRF 9, 9 is defined as "COUNTER," and written CLRF COUNTER. Clearly, the purpose of this line of code is much more evident than 69h. As the program is written, each line of code has a sequential address. To call a subroutine or jump to a new program location, you need to know the address of those locations. The assembler keeps track of this. All you have to do is assign a label to the beginning of your routine and reference the label in the call or jump statement. The Microchip assembler is called MPASM. It will support all variations of the PIC® 16CXX and PIC® 17CXX.

C language takes the idea of the assembler and goes one step further. As the assembler frees you from the raw coding of the chip, C frees you from writing each line of Assembly language. It may take a dozen lines of Assembly language to perform one C language command. C also allocates memory for variables, relieving the programmer from keeping track of every memory location. The C compiler from Microchip is called MP-C.

In the same vein, there also are Basic compilers for the PIC® line. A single Basic command such as *write character to serial port* may take a hundred lines of Assembly language to produce. At this time, Microchip does not offer a Basic compiler. Each higher-level language is easier to write but at the sacrifice of speed and program (compiled code) size. An application written in Basic could be twice as slow and twice as large as the same program in Assembly language. The reasons for this are complex and not suitable for discussion in this book. Suffice it to say, you will be learning and using an assembler to gain maximum speed and efficiency.

A simulator is a software program that runs on your average computer. The program loads the firmware written for the PIC® and simulates the PIC® running the code. You can *single-step* through each line of code, watching the values in each register change. It is possible to insert *break points* in the code so that large chunks can be run quickly, stopping at relevant times for analysis. The simulator is a useful tool at times, but is often limited by the need to interface with the real world. Since most microcontroller applications deal with controlling something, feedback or sensor input is required for the program to function properly. You can single-step the code, inserting the proper input to run correctly. This can be, and is, tedious.

A better solution is the emulator. One end of the emulator fits on the application PCB in the same position as the PIC®, while the other end is connected to a PC. Now you are running the firmware code, emulating all the inputs and outputs of the PIC®, from the PC. As with the simulator, you can single-step the code, checking each register as you go. This is a very powerful tool for microcontroller development. Emulators are usually expensive, often costing more than six hundred dollars. The PIC® emulator from Microchip is the PICMASTER®.

Third Party Support

While Microchip is the company that manufactures PIC® microcontrollers, this does not mean that they produce the only support products for the PIC®. We will now examine a few of the many firms that offer PIC® third party support items.

Micro Engineering Labs has one of the lowest cost programmers on the market today. The EPIC programmer can handle the PIC® 16C61, 62x, 64, 65, 71, 73, 74, and 84. The assembled version costs around $59.95, and a bare-board version goes for only $34.95. At these prices, there are no excuses for not adding a programmer to your work bench. Adapters are required for 28- and 40-pin PICs®. They also offer a line of PIC® prototype boards. These PCBs have the circuitry for the PIC®, voltage regulators, capacitors, and oscillator function. The rest of the board has hundreds of plated-through holes (on a .1 inch matrix) for your custom parts. A quick solution for prototypes.

J&M Microtek, Inc., produces a number of development tools for a variety of microcontrollers. For the PIC® line, they offer an all-in-one device, the JM-PICE. It can program 16C5x, 16C64, 16C71, 16C84, and 17C42. Software includes the assembler, disassembler, and simulator. Surprisingly, this device also functions as an in-circuit emulator for 16C5x. Finally, it can also perform instruction level emulation for the 16C71, 16C84 and 17C42.

Custom Computer Services has written an excellent C compiler for the PIC® line. In addition to most standard C commands, this compiler has dozens of specialized instructions that recognize the hardware nature of the PIC®. It also supports in-line Assembly code. To illustrate the power of the C compiler, let's look at the implementation of a serial port in a PIC®. To initiate the port, the compiler command is *#USE RS232(BAUD=9600, XMIT=PIN_1,RCV=PIN_2);*. To send an entire sentence the command would be *puts("Hello");*. Compare this to the RS232 examples in Chapter 8. What a savings in time and effort!

Last, but certainly not least, is Parallax, Inc. They offer a full line of PIC® support products including programmers, emulators, simulators, Assembly language, and BASIC stamp modules. A BASIC stamp module is a programmed PIC® (programmed with a Basic interpreter),

EEPROM for program storage (of the Basic program), oscillator and I/O lines. The purpose of this device is quick development and small production runs. Since the program is downloaded and stored in EEPROM, plus written in simple Basic language, the time required to write and debug the program can be under an hour!

The Parallax programmer can be purchased for around $99.00. It includes sockets for 18- and 28-pin PICs®. The Parallax assembler is a valuable tool. It is not a one-for-one correspondence with the native PIC® instruction code, but rather a super-set of it. To compare, the Microchip instruction set has 69 instructions while the Parallax assembler has 119 instructions. Where did the extra instructions come from? They are pseudo-instructions composed of multiple real instructions. Let's look at one example, the Parallax instruction CSNE (compare and skip if not equal). The line in Parallax assembly is *CSNE fr1,fr2*. The equivilant code in the Microchip assembler is *MOVF fr2,0 , SUBWF fr1,0 , BTFSC 3,2*. The Parallax code is much clearer and easier to read. For this reason, the projects in this book are written in Parallax Assembly language. The next chapter goes into much greater depth on programming, and we will see many examples of code.

CHAPTER 6
PIC® Firmware

As you are now aware, firmware is the program that tells the microcontroller what to do. With only 33 instructions the PIC® family of controllers can do an almost unlimited number of tasks. The instructions are intimately tied to the architecture of the chip, hence the prior lessons on PIC® hardware and how it works. We will now delve into what each instruction does, its mnemonic, timing, and flags affected.

We will look at three different ways to program the PIC®. First are the raw and basic commands of the PIC® itself. While you will not be programming in machine language, understanding the most elementary programming form of the PIC® will be of lasting and useful benefit. Second is a slightly higher language, the Parallax Assembly code. The PIC® projects later in the book will be written in Parallax. We will also look briefly at a C language compiler and what it has to offer in a micro-controller environment. Finally, you shall see a comparison between the different languages on the same application.

Instruction Set

In the following description, each instruction starts with its mnemonic and name. The hex code is either absolute or symbolic if it contains a lower case f, k, or b. The f (file register), k (constant or literal), or b (bit location) represents a range of possible values. Hex 02f could be 021, 022, 023, etc., depending on which register the instruction will work with. Then comes a short description of how the instruction functions. Last is the amount of time for each execution, and which internal flags are affected. Remember, one cycle is 1/4 the oscillator frequency. Flags are Z-*result was a zero*, C-*carry*, DC-*nibble carry*, TO-*watchdog timer time out*, and PD for *power down*:

Name:	NOP - No operation
Hex:	000
Operation:	Does nothing. Good for adjusting critical timing functions.
Cycles, Flags:	1, None

Name: MOVWF - Move W to f
Hex: 02f
Operation: The value in the work register will move to the specified register.
Cycles, Flags: 1, None

Name: CLRW - Clear W
Hex: 040
Operation: Clear the work register.
Cycles, Flags: 1, Z

Name: CLRF - Clear f
Hex: 06f
Operation: Clear the specified register.
Cycles, Flags: 1, Z

Name: SUBWF - Subtract W from f
Hex: 08f
Operation: Subtracts the work register from the specified register.
Cycles, Flags: 1, C, DC, Z

Name: DECF - Decrement f
Hex: 0Cf
Operation: One is subtracted from the specified register, and the result is moved into either the specified register or the work register.
Cycles, Flags: 1, Z

Name: IORWF - Inclusive OR W and F
Hex: 10f
Operation: The work register is ORed with the specified register, and the result is moved into either the specified register or the work register.
Cycles, Flags: 1, Z

Name: ANDWF - AND W and f
Hex: 14f
Operation: The work register is ANDed with the specified register, and the result is moved into either the specified register or the work register.
Cycles, Flags: 1, Z

Name: XORWF - Exclusive OR W and f
Hex: 18f
Operation: The work register is exclusive ORed with the specified register, and the result is moved into either the specified register or the work register.
Cycles, Flags: 1, Z

Name: ADDWF - Add W and f
Hex: 1Cf
Operation: The specified register is added with the work register, and the result is moved into either the specified register or the work register.
Cycles, Flags: 1, C, DC, Z

Name: MOVF - Move f
Hex: 20f
Operation: The specified register is moved into the work register.
Cycles, Flags: 1, Z

Name: COMF - Complement f
Hex: 24f
Operation: The complement of the specified register is moved into the work register or back into the specified register.
Cycles, Flags: 1, Z

Name: INCF - Increment f
Hex: 28f
Operation: One is added to the specified register, and the result is moved either into the specified register or the work register.
Cycles, Flags: 1, Z

Name: DECFSZ - Decrement f, Skip if 0
Hex: 2Cf
Operation: One is subtracted from the specified register, and the result is moved either into the specified register or the work register. If the result is 0, then the next instruction is skipped.
Cycles, Flags: 1 or 2 on skip, None

Name: RRF - Rotate right f
Hex: 30f
Operation: The specified register is rotated right one bit, and the result is moved either into the specified register or the work register.
Cycles, Flags: 1, C

Name: RLF - Rotate left f
Hex: 34f
Operation: The specified register is rotated left one bit, and the result is moved either into the specified register or the work register.
Cycles, Flags: 1, C

Name: SWAPF - Swap halves f
Hex: 38f
Operation: The upper and lower nibble of the specified register are exchanged. The result is moved into the specified register or the work register.
Cycles, Flags: 1, None

Name: INCFSZ - Increment f, Skip if 0
Hex: 3Cf
Operation: Add one to the specified register, and skip the next instruction if the specified register is equal to 0.
Cycles, Flags: 1 or 2 if skip, None

Name: BCF - Bit Clear f
Hex: 4bf
Operation: The specified bit of the specified register is set to 0.
Cycles, Flags: 1, None

Name: BSF - Bit Set f
Hex: 5bf
Operation: The specified bit of the specified register is set to 1.
Cycles, Flags: 1, None

Name: BTFSC - Bit Test f, Skip if Clear
Hex: 6bf
Operation: If the specified bit of the specified register is 0, then skip the next instruction.
Cycles, Flags: 1 or 2 if skip, None

Name: BTFSS - Bit Test f, Skip if Set
Hex: 7bf
Operation: If the specified bit of the specified register is 1, then skip the next instruction.
Cycles, Flags: 1, None

Name: OPTION - Load OPTION register
Hex: 002
Operation: The work register is moved into the option register.
Cycles, Flags: 1, None

Name: SLEEP - Go into standby mode
Hex: 003
Operation: 0 is moved into the watchdog timer, and the oscillator stops.
Cycles, Flags: 1, TO, PD

Name: CLRWDT - Clear Watchdog Timer
Hex: 004
Operation: 0 is moved into the watchdog timer.
Cycles, Flags: 1, TO, PD

Name: TRIS - Tristate port f
Hex: 00f
Operation: The work register is moved into the control register of the specified register. This sets which lines are outputs or inputs.
Cycles, Flags: 1, None

Name: RETLW - Return, place Literal in W
Hex: 8kk
Operation: Exit from a subroutine. Pop the stack to the program counter, and load the work register with the specified constant.
Cycles, Flags: 2, None

Name: CALL - Call subroutine
Hex: 9kk
Operation: Jumps to subroutine. The current program counter, plus one is saved on the stack and the specified address is moved to the program counter.
Cycles, Flags: 2, None

Name: GOTO - Go To Address
Hex: Akk
Operation: Specified address is moved to the program counter.
Cycles, Flags: 2, None

Name: MOVLW - Move Literal to W
Hex: Ckk
Operation: The specified constant is move to the work register.
Cycles, Flags: 1, None

Name: IORLW - Inclusive OR Literal and W
Hex: Dkk
Operation: The specified constant is ORed with the work register.
Cycles, Flags: 1, Z

Name: ANDLW - AND Literal and W
Hex: Ekk
Operation: The specified constant is ANDed with the work register.
Cycles, Flags: 1, Z

Name: XORLW - Exclusive OR Literal and W
Hex: Fkk
Operation: The specified constant is exclusive ORed with the work register.
Cycles, Flags: 1, Z

Well, that's the entire PIC® instruction set! With these few and simple commands, you can create powerful and intelligent control applications. You may have a moist brow after considering every f, b, and k, but take heart; very few people write code like F23, 1C3, 512, etc. That's where Assembly language comes to the rescue.

Assembly Language

Assembly language takes care of routine bookkeeping so you can concentrate on program flow. As you saw in the previous section, each PIC® instruction can have many variations depending on registers address, bit address, or constant. A typical Assembly language line would be *GOTO count_again*—this would be translated into 923 for the PIC®. Obviously, the Assembly language is much clearer. Of course, you have to tell the assembler where *count_again* is in the program. This and other protocols are the format of the language. We will examine this and see many examples later in the book.

The assembler we will study is from Parallax, Inc., and is called PASM. Parallax expanded the number of instructions by lumping four native instructions into a new one. This created a range of very useful pseudo-instructions without the usual overhead of a high level compiler. An example of this would be the CJE command. In Parallax you would write:

  CJE f1,f2,addr9   (compare register 1 to register 2 and jump to addr9 if equal)

The equivalent code in native Assembly would be:

  MOVF f2,0   (move register 2 to the work register)
  SUBWF f1,0   (subtract register 1 from the work register)
  BTFSC 3,2   (test the 0 flag, skip if clear (subtraction not 0))
  GOTO addr9   (jump to addr9)

All of the native instructions are in the Parallax assembler as well as their new instructions. Let's look at some of the more interesting Parallax commands. Keep in mind that the jumps to addresses are 9 bits, and therefore limited to 512 bytes (1 page). A LJMP (long jump) can jump to any location in program memory:

Name: ADDB - Add bit into register.
Syntax: ADDB register, bit address
Example: addb count_high,c
Operation: If the specified bit is set, then one is added to the specified register. This can be used in multiple byte addition. When the lower bytes are added, they will set the carry bit if an overflow occurred. This instruction can test the carry bit and then increment the next higher byte in the number.

Name: CJA - Compare register to literal and jump if above.
Syntax: CJA register, #literal, address
Example: CJA buf1,#34,function2
Operation: If the specified register is larger than the supplied literal, a jump is made to the specified address.

Name: CJA - Compare register1 to register2 and jump if above.
Syntax: CJA register1, register2, address
Example: cja count,limit,exit
Operation: If register1 is greater than register2, then a jump is made to the specified address.

Name: CJAE - Compare register to literal and jump if above or equal.
Syntax: CJAE register, #literal, address
Example: cjae input,#10,beep
Operation: If the register is equal to or larger than the supplied literal, then a jump is made to the specified address.

Name: CJAE - Compare register1 to register2 and jump if above or equal.
Syntax: CJAE register1, register2, address
Example: cjae new_num,old_max,set_new
Operation: If register1 is equal to or larger than register2, then jump to the specified address.

Name: CJB - Compare register to literal and jump if below.
Syntax: CJB register, #literal, address
Example: cjb reps,#25,not_done
Operation: If the specified register is less than the supplied literal, then a jump is made to the specified address.

Name: CJB - Compare register1 to register2 and jump if below.
Syntax: CJB register1, register2, address
Example: cjb input,low,new_low
Operation: If register1 is less than register2, then a jump is made to the specified address.

Name: CJE - Compare register1 to register2 and jump if equal.
Syntax: CJE register1, register2, address
Example: cje in_char,trigger,led_on
Operation: If register1 is equal to register2, then a jump is made to the specified address.

Name: CJNE - Compare register to literal and jump if not equal.
Syntax: CJNE register, #literal, address
Example: cjne count,#10,loop_again
Operation: If the register is not equal to the supplied literal, then a jump is made to the specified address.

Name: CSA - Compare register to literal and skip if above.
Syntax: CSA register,#50
Example: csa buf1,#25
Operation: If the register is larger than the supplied literal, then the next instruction is skipped. The next instruction must be a single-byte type instruction (such as the native instruction set) and not a Parallax pseudo-instruction.

Name: DJNZ - Decrement register and jump if not 0.
Syntax: DJNZ register, address
Example: djnz rep_num,one_more_time
Operation: One is subtracted from the register, and if the register is not zero, then a jump is made to the specified register.

Name: JNB - Jump if not bit.
Syntax: JNB bit address, address
Example: jnb flag.2,read_port
Operation: If the specified bit is 0, then a jump is made to the specified address.

Name: LCALL - Long call.
Syntax: LCALL address
Example: lcall send_char
Operation: This instruction allows subroutine calls to any page in the PIC® by setting the page select bits. The call still must be to the first 256 bytes of the page. Return from the routine does not automatically reset the page bits. This must be done with the LSET command.

Name: LJMP - Long jump.
Syntax: LJMP address
Example: ljmp far_away
Operation: Jump to any address in the PIC®.

Name: LSET - Long set.
Syntax: LSET address
Example: lset start
Operation: Resets the page select bits to the page that the specified address resides in. LCALL, LJMP, and LSET are only useful on PIC's with more than one page, such as the 16C56 and 16C57.

Name: MOVB - Move bit2 to bit1.
Syntax: MOVB bit address1, bit address2
Example: movb flag.3,portA.1
Operation: Bit 1 is changed to 1 or 0 depending on the value of bit2.

The previous commands are just a small portion of the available instructions in the Parallax assembler. If you purchase the assembler, then the Parallax manual (about two hundred pages) gives a full description of all instructions. In the interest of brevity, and since the upcoming projects are written in Parallax Assembly, *Table 6-1* shows all Parallax commands and syntax. By now, most will be self explanatory. If any instructions appear vague as to function, please consult the Parallax manual.

C Compiler

The C language is arguably the most widely used language on small computers today. Operating systems (DOS, Windows, Unix) and major applications (database, spreadsheet, word processor) are written in C. C has the great advantage of being highly transportable; this means the same code

MNEMONIC	NAME	SYNTAX
ADD	Add literal into register	ADD fr,#literal
ADD	Add register2 into register1	ADD fr1,fr2
ADD	Add work register into register	ADD fr,W
ADD	Add register into work register	ADD W,fr
ADDB	Add bit into register	ADDB fr,bit
AND	AND literal into fr	AND fr,#literal
AND	AND register2 into register1	AND fr1,fr2
AND	AND work register into register	AND fr,W
AND	AND literal into work register	AND W,#literal
AND	AND register into work register	AND W,fr
CALL	Call subroutine	CALL addr8
CJA	Compare register to literal and jump if above	CJA fr,#literal,addr9
CJA	Compare register1 to register2 and jump if above	CJA fr1,fr2,addr9
CJAE	Compare register1 to literal and jump if above or equal	CJAE fr,#literal,addr9
CJAE	Compare register1 to register2 and jump if above or equal	CJAE fr1,fr2,addr9
CJB	Compare register to literal and jump if below	CJB fr,#literal,addr9
CJB	Compare register1 to register2 and jump if below	CJB fr1,fr2,addr9
CJBE	Compare register to literal and jump if below or equal	CJBE fr,#literal,addr9
CJBE	Compare register1 to register2 and jump if below or equal	CJBE fr1,fr2,addr9
CJE	Compare register to literal and jump if equal	CJE fr,#literal,addr9
CJE	Compare register1 to register2 and jump if equal	CJE fr1,fr2,addr9
CJNE	Compare register to literal and jump if not equal	CJNE fr,#literal,addr9
CJNE	Compare register1 to register2 and jump if not equal	CJNE fr1,fr2,addr9
CLC	Clear carry	CLC
CLR	Clear register	CLR fr
CLR	Clear work register	CLR W
CLR	Clear the watchdog timer	CLR WDT
CLRB	Clear bit	CLR bit
CLZ	Clear zero bit	CLZ
CSA	Compare fr to literal and skip if above	CSA fr,#literal

Continued...

CSA	Compare register1 to register2 and skip if above	CSA fr1,fr2
CSAE	Compare register to literal and skip if above or equal	CSAE fr,#literal
CSAE	Compare register1 to register2 and skip if above or equal	CSAE fr1,fr2
CSB	Compare register to literal and skip if below	CSB fr,#literal
CSB	Compare register1 to register2 and skip if below	CSB fr1,fr2
CSBE	Compare register to literal and skip if below or equal	CSBE fr,#literal
CSBE	Compare register1 to register2 and skip if below or equal	CSBE fr1,fr2
CSE	Compare register to literal and skip if equal	CSE fr,#literal
CSE	Compare register1 to register2 and skip if equal	CSE fr1,fr2
CSNE	Compare register to literal and skip if not equal	CSNE fr,#literal
CSNE	Compare register1 to register2 and skip if not equal	CSNE fr1,fr2
DEC	Decrement register	DEC fr
DECSZ	Decrement register and skip if zero	DECSZ fr
DJNZ	Decrement register and jump if not zero	DJNZ fr,addr9
IJNZ	Increment register and jump if not zero	IJNZ fr,addr9
INC	Increment register	INC fr
INCSZ	Increment register and skip if zero	INCSZ register
JB	Jump if bit is set (1)	JB bit,addr9
JC	Jump if carry bit is set (1)	JC addr9
JMP	Jump to address	JMP addr9
JMP	Jump to PC+W (program counter + work register)	JMP PC+W
JMP	Jump to W (work register moved into program counter)	JMP W
JNB	Jump if not bit (bit is clear 0)	JNB bit,addr9
JNC	Jump if not carry (carry bit is clear 0)	JNC addr9
JNZ	Jump if not zero (zero bit is clear 0)	JNZ addr9
JZ	Jump if zero (zero bit is set 1)	JZ addr9
LCALL	Long call	LCALL addr11
LJMP	Long jump	LJMP addr11
LSET	Long set	LSET addr11
MOV	Move literal into register	MOV fr,#literal
MOV	Move register2 into register1	MOV fr1,fr2
MOV	Move work register into register	MOV fr,W

Continued...

MOV	Move literal into OPTION register	MOV OPTION,#literal
MOV	Move register into OPTION register	MOV OPTION,fr
MOV	Move work register into OPTION register	MOV OPTION,W
MOV	Move literal into I/O port control register	MOV !port_fr,#literal
MOV	Move register into I/O port control register	MOV !port_fr,fr
MOV	Move work register into I/O port control register	MOV !port_fr,W
MOV	Move literal into work register	MOV W,#literal
MOV	Move register into work register	MOV W,fr
MOV	Move not register into work register	MOV W,/fr
MOV	Move (register - work register) into work register	MOV W,fr-W
MOV	Move the incremented value of register into work register	MOV W,++fr
MOV	Move the decremented value of register into work register	MOV W,--fr
MOV	Move the left-rotated value of register into work register	MOV W,<<fr
MOV	Move the right-rotated value of register into work register	MOV W,>>fr
MOV	Move the nibble-swapped value of register into work register	MOV W,<>fr
MOVB	Move bit2 to bit1	MOVB bit1,bit2
MOVB	Move not bit2 to bit1	MOVB bit1,/bit2
MOVSZ	Move the incremented value of register into work register and skip the next instruction if zero	MOVSZ W,++fr
MOVSZ	Move the decremented value of register into work register and skip the next instruction if zero	MOVSZ W,--fr
NEG	Negate register (two's complement)	NEG fr
NOP	No operation	NOP
NOT	Not register (one's complement)	NOT fr
NOT	Not work register	NOT W
OR	OR literal into register	OR fr,#literal
OR	OR register2 into register1	OR fr1,fr2
OR	OR work register into register	OR fr,W
OR	OR literal into work register	OR W,#literal
OR	OR register into work register	OR W,fr
RET	Return from subroutine	RET

Continued...

RETW	Assemble RET's which load work register with literal data (can build a data table with this command)	RETW literal1,literal2, literal3,....
RL	Rotate left register	RL fr
RR	Rotate right register	RR fr
SB	Skip if bit is set (1)	SB bit
SC	Skip if carry is set (1)	SC
SETB	Set bit (1)	SETB bit
SKIP	Skip next instruction	SKIP
SLEEP	Enter sleep mode	SLEEP
SNB	Skip if not bit (bit is 0)	SND bit
SNC	Skip if not carry (carry bit is 0)	SNC
SNZ	Skip if not zero (zero bit is 0)	SNZ
STC	Set carry bit (1)	STC
STZ	Set zero bit (1)	STZ
SUB	Subtract literal from register	SUB fr,#literal
SUB	Subtract register2 from register1	SUB fr1,fr2
SUB	Subtract work register from register	SUB fr,W
SUBB	Subtract bit from register	SUBB fr,bit
SWAP	Swap nibbles in register	SWAP fr
SZ	Skip if zero (zero bit is 1)	SZ
TEST	Test register for zero	TEST fr
TEST	Test work register for zero	TEST W
XOR	XOR literal into register	XOR fr,#literal
XOR	XOR register2 into register1	XOR fr1,fr2
XOR	XOR work register into register	XOR fr,W
XOR	XOR literal into work register	XOR W,#literal
XOR	XOR register into work register	XOR W,fr

Table 6-1. *Parallax PIC® assembler commands.*
(Continued from Pages 71-74.)

Information reprinted with permission from
PIC® 16C5x Development Tools - Assembler, Programmer, Emulator, ©Parallax, Inc.

means the same code written to run on a PC can be compiled to run on a Unix workstation. Most processors and microcontrollers have a C compiler available.

The advantages to C are English-like command statements, memory management, and simplified interrupt procedures. Let's look at the following code sample:

if (n<10) putchar(n+48);
printf("Press any key to continue");

Without ever opening a book on C, it would be obvious to anyone the first line "does something" if the value in *n* is less than 10. With the knowledge that *putchar* sends a byte to an I/O port (serial, parallel, CRT, etc.), you can see the total line sends *n*+48 to an I/O port if *n* is less than 10. The only specialized knowledge to understand in the second line is that *printf* sends an entire text string to an I/O port. Of course, while learning the whole C language is a sizable undertaking that should not be trivialized, the point is that C language is much easier to read and write than Assembly language.

With Assembly, the programmer has to keep track of where variables and data are in memory. Moving bytes to the wrong addresses would be fatal. With C language, the compiler keeps track of variables and data. In the previous C code sample, *n* would be declared at the start of the program as being an integer variable. C assigns a memory location for *n*, and no other data can be written to it. Since a large program can contain hundreds of variables and matrixes, memory management is invaluable.

In the previous discussion on interrupts, you discovered the hazards when the interrupt is executed. Many of the specialized registers can hold information which must remain intact upon return from the interrupt. A number of C compilers (not all of them) will automatically save the SFRs (by pushing them on the stack) and then retrieve them upon return from interrupt.

The disadvantages are speed and efficiency of code. Compiled C code is often larger than the same program written in Assembly. This often causes the code to run slower. Most programmers feel this is offset by the ease of writing C compared to Assembly. On a large project, writing in C can be three or more times faster.

C language cannot be taught here, but hopefully you have received a taste of what C can do. To learn more, the classic *The C Language* by Brian W. Kernighan and Dennis M. Ritchie (creators of the language) is highly recommended.   A very good C compiler for the PIC® is PCB by Custom Computer Services.

Comparison

Let's do a quick comparison between PIC® native Assembly, Parallax Assembly, and PCB C compiler for the same project. This is a very simple example, creating a 1Kz square wave on an output pin. The PIC® is the 16C54 with a crystal operating at 20Mhz. Keep in mind that this example is simple and that the power of C not fairly shown:

PIC® ASSEMBLY

```
; 1 Khz signal
; variables and equates
output = ra.1
count_low = 7
count_high = 8
        org 0
wait                    ;500 us wait
        movlw 155
        movwf count_low
        movlw 1
        movwf count_hi
wait_loop
        decfsz count_low,F
        goto   wait_loop
        decfsz count_hi,F
        goto   wait_loop
        retlw   0


; Reset address
        reset start

start
        movlw 0
        tris    5       ;set RA all output
        movlw 0
```

```
        tris   6         ;set RB all output
        loop
        bsf    output ;set output high
        call   wait    ;wait 500us
        bcf    output ;set output low
        call   wait    ;wait 500us
        goto   loop    ;endless loop

    ;end
```

PARALLAX ASSEMBLY

```
    ; 1 Khz signal
    ; Variables and equates
    output = ra.1
        org  7          ;start address of file register
    count_low   ds   1
    count_hi    ds   1
        org 0           ;reset code origin
    ; Device data
        device pic16c54,hs_osc,wdt_off,protect_off
    ;
    wait                  ;500 us wait
        mov    count_low,#155
        mov    count_hi,#1
    wait_loop
        djnz   count_low,wait_loop
        djnz   count_hi,wait_loop
        ret
    ;
    ;reset address
    ;
        reset  start
    ;
    start
        mov    !ra,#0  ;set RA all output
        mov    !rb,#0  ;set RB all output
    loop
```

```
            setb    output ;output high
            call    wait   ;wait 500us
            clrb    output ;output low
            call    wait   ;wait 500us
            jmp     loop   ;endless loop
    ;
    ;end
    ;
```

PCB C LANGUAGE

```
    /* 1 Khz signal */
    #include <PIC16C54.H>

    #use Delay(Clock=20000000)

    main() {
          while (TRUE) {
                  output_high(PIN_8);
                  delay_us(500);
                  output_low(PIN_8);
                  delay_us(500);
                          }
          }
```

Both the PIC® and Parallax Assembly examples compile to 19 bytes. The C language example compiles to 35 bytes. In these examples, all of the firmware runs at the same speed. It is up to your personal preference as to which language to use. They can all create the same functioning code.

CHAPTER 7
The First Project:
Lights, Action!*

The best way to learn is through practice and experience. The previous chapters introduced you to the various components of a microcontroller. To create an application, you must successfully integrate those components into a working whole. There are three basic steps toward creating a microcontroller application. First, you must understand exactly what the controller is expected to do: i.e., the objective. Second, you need to design the electronic circuity that enable the controller to perform the job: i.e., the hardware. Lastly, you need to write the microcontroller program that accomplishes the objective and provides ease of user interface: i.e., the firmware.

This initial project allows you to start at a beginner's level. The objective is to move a light in a circle based on input from a random number generator. This chapter was originally written as an article for *Popular Electronics* under the title "Build the PK Tester." It was published in the May 1995 issue of *P.E.*, and again in the Fall 1995 issue of *Electronics Hobbyists Handbook*. The configuration of the unit allows for scientific testing for the psychokinetic (PK) ability to remotely influence a random number generator. If you're not into ESP, the device can be easily reprogrammed to be a roulette wheel, light show, or other interesting projects. At the end of the chapter, we will examine important features of the firmware, which we will also do with each of the five projects. Hopefully, this will give you a better handle on how to write your own firmware. In an appendix at the end of the book, the complete firmware for all of the projects are given.

Objective

Psychokinesis is the supposed ability of being able to move objects or influence events with one's mind. Over the years, many scientific investigations have gathered surprising evidence to support a belief in PK, but there is a continued skepticism among the general population. If

*This chapter was originally written as the article, "Build the PK Tester," in the May, 1995 issue of *Popular Electronics*. Reprinted with permission from *Popular Electronics* Magazine, May 1995 issue. ©Copyright Gernsback Publications, Inc., 1995

you would like to settle that issue for yourself and your friends, build the PK tester described in this chapter.

The project duplicates a device created by the German physicist Helmut Schmidt. In 1969, Schmidt was working for the Boeing Company, which allowed him the time and resources to do PK and ESP research. With his traditional physics training, he believed psychic powers could not exist. However, being open-minded, Schmidt decided to design an experiment that would scientifically attempt to resolve the mind-over-matter question.

The heart of Schmidt's device was a random-number generator (RNG). To make the RNG, Schmidt used a radioactive substance called Strontium-90, which created a random strobe due to its erratic decay. This caused a sample to be taken of a 50%-duty-cycle square wave at random intervals. (*Figure 7-1*). The result was a totally random series of lows and highs (zeros and ones) that would cancel out over time.

Schmidt linked the output of his RNG to a box with 10 lights on it, arranged in a circle. Only one light would be lit at a time; therefore, the light would give the appearance of moving counterclockwise or clockwise depending on the RNG state. When the device was not being "influenced" by PK, it would drift in one direction or the other, but over time, its movement should be statistically neutral.

A person whose PK abilities were being tested was asked to "think" the lights in either a clockwise or counterclockwise direction. It is interesting to note that the people tested were not aware of the underlying method or electronics used to create the light's movement. Schmidt's results were startling: some people influenced the motion of the light by odds of over 10,000 to 1.

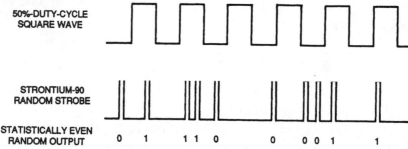

Figure 7-1. *Here's a sample of the statistically-even output produced by Schmidt's random number generator. A similar output is generated by the PK tester using a nonradioactive method.*

In his day, Schmidt used some very advanced equipment to carry out his experiments. However, Strontium-90 is a radioactive isotope that many scientists believe to be dangerous. For that reason, a different method is used to create the random strobe in this project (discussed later). However, developing an RNG is only part of the process of recreating Schmidt's experiment. A method is also needed for translating the random output into directional movement, and for keeping track of the direction of that movement.

The solution to the problem is a Microchip Technology PIC® 16C55 microcontroller. The PIC's® RISC-like architecture combined with its top clock of 20 MHz, allows it to process 5 million instructions per second (MIPS), making it one of the fastest microcontrollers around. The PIC® used in the PK tester is widely available from a number of national distributors.

Circuit Description

The schematic for the PK tester is shown in *Figure 7-2*. Power is provided by a 12-volt DC wall adapter, which plugs into power-jack J1. It is used instead of batteries because, to guarantee the PK tester's randomness, the circuit should be tested for many hours or even days at a stretch. A regulated 10 volts is needed by the noise circuit, provided by U4, an LM317 adjustable voltage regulator. High ripple rejection is accomplished by R11 and C10. The rest of the circuit runs off a regulated 5 volts, produced by U5, a 7805.

Resistor R2 and capacitor C1 provide the RC timing for the PIC® (U1), giving it a 78 μs clock. The cathodes of LED1 - LED16 are directly connected to the PIC®. The anodes of the LEDs are connected in common through R1, a 220-ohm resistor, to +5 volts. During normal use only one LED is on at a time, bypassing the need for separate current-limiting resistors on each LED. Two push-button switches, S1 and S2, are used to initiate the STATUS/TEST and RESET functions described later.

The random strobe is produced this way: transistor Q1's emitter-to-base junction is reverse-biased over the breakdown point. This type of configuration produces random noise that is then amplified by Q2. The resulting output is fed into U2, an LM311 comparator, and comes out as a clean, TTL logic-level high or low signal.

The other signal needed to reproduce Schmidt's experiment is a square wave with a 50% duty cycle. It is created by U3, a 555 timer. Diodes D1 and D2 are used to generate the separate timing paths necessary for a precision 50% duty cycle. Also, to ensure that the square wave has a perfect 50% duty cycle, potentiometer R5 should be properly adjusted (more on this

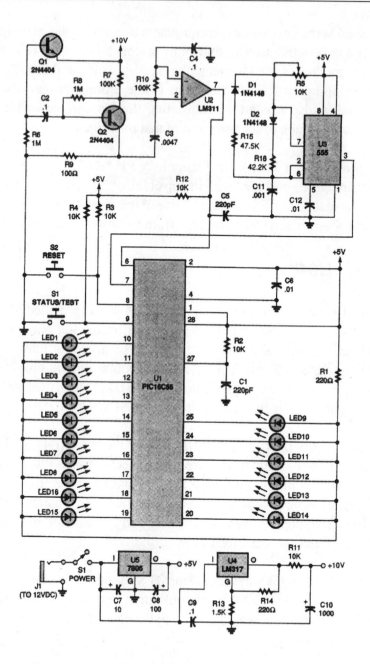

Figure 7-2. *This circuit generates a random output that is translated into LED "movement" by a preprogrammed PIC® 16C55 microcontroller, U1. This PIC® also senses and records the bias of the LED's movement.*

later). The combination of the square wave plus the random strobe equals a random but statistically neutral series of highs and lows.

The output of the highs and lows is fed to pins 6 and 7 of U1. A section of the firmware then translates the signals into LED "movement." Each individual movement is also recorded by the PIC®, and the total number of movements are tabulated for display once the STATUS/ TEST button is pressed.

Construction

The construction technique used is not critical; a wired prototype would work. However, if you would like to build the circuit on a PCB, a pattern is shown in *Figure 7-3*. Other parts for the PK tester can be readily acquired from hobbyist sources like Radio Shack or Digi-Key.

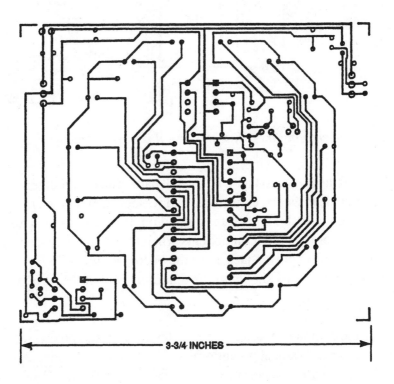

3-3/4 INCHES

Figure 7-3. *Use this template to etch your own PC board.*

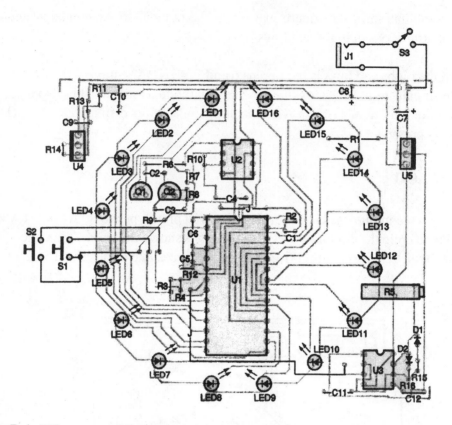

Figure 7-4. *When assembling the PK tester, use this parts-placement diagram as a guide.*

If you decide to build the project on a PCB, use the parts-placement diagram in *Figure 7-4* as a guide. Mount all of the resistors, capacitors, and IC sockets first; then insert the ICs in their sockets. Last, mount the LEDs so that they stand on 3/4-inch leads. This will ensure that there is enough room for the other components when the circuit is placed in the case. Wire the ground connection from power-jack J1 to the board, as shown in *Figure 7-4*, then make the other off-board connections shown in the diagram. Mount the switches and the jack to the case.

Figure 7-5 shows the complete board for the PK tester. Before you can mount the PC board in the case, you will have to adjust potentiometer R5 so that U3 produces a 50%-duty-cycle square wave. The best way to accomplish this is by using a scope. However, if you do not have a scope handy, you could adjust R5 to an approximately "halfway" setting, and then use trial and error to determine if the output of the PK tester is statistically neutral.

Figure 7-5. *The complete board for the PK tester.*
Notice the height of the LEDs for placement into a case.

With R5 adjusted, you can complete the assembly of the unit. Drill 16 holes on the top panel of the project case to match the pattern of the LEDs. Insert an LED lens cap into each of the holes, then lift the board and press the LEDs into their respective caps. When this is done, the PC board will be held firmly under the panel.

Operation

The finished unit has two push-button switches (S1 and S2). If the STATUS/TEST button (S1) is depressed when power-switch S3 is turned on, the tester will go into a self-test mode. It will take 480 samples to see if the RNG is operating correctly. If it is, all of the LEDs will flash and the unit will then proceed into its normal operation mode. If there is a problem with the circuit, all of the LEDs will continue flashing, indicating an error, until the unit is powered down. Check the calibration of R5 if this occurs.

Once in the normal operation mode, the PIC® keeps track of LED "movements" to the left or right. To find out if one direction has been coming up more frequently, hold down the STA-TUS/TEST button. If the top two LEDs light up, then there is no excess movement. If one or more light to the left or right, it means the unit is biased in the indicated direction. To determine the *maximum* number of moves that the unit is biased in that direction, use:

$$n \times 16$$

where *n* is the number of LEDs lit. The system of LEDs used in the PK tester can only indicate the left or right bias in increments of 16 moves. For that reason, the *minimum* number of moves is 15 less than the maximum number. For example, if three LEDs are lit to the right, then movement to the right exceeds normal by 33 to 48 moves. As a guide, a reading of seven LEDs to the left or right indicates an extreme bias in that direction. When you have finished taking a reading, you can either let go of the STATUS/TEST button and resume testing, or clear the "movement" counters and restart the PK tester by pressing S2.

Leave the box on for 24 hours and verify its randomness by pressing STATUS/TEST. Then, have a test subject concentrate on the PK tester. Have him or her "will" the movement of the LED either clockwise or counterclockwise. After about ten minutes, verify whether the LED movements are neutral by again pressing the STATUS/TEST button. Should the LED movement be biased in one direction, have the subject continue his or her "willing" so that you can see if the bias is coincidental. After another ten minutes, the status response should show an even greater bias if the RNG actually is being influenced. Next, press the RESET button and ask the subject to concentrate in the reverse direction. If the bias follows the desired direction after several reversals, your subject most likely has a notable level of PK ability.

Firmware

As with all of the projects in this book, we will review highlights of the firmware. Since this is the first project, we will also examine the general style and format of the program. First and foremost, put plenty of comments in your program! This has been said in a thousand books on programming, and it is worth repeating. If anyone has to modify your code later, the comments will be of great value in helping someone else understand the code. "Ah," you say, "No one else will work on *my* code!" Don't be too sure of that. Besides, the comments will help you understand the code during the initial writing and later rewriting of the code. In Assembly language, a comment is created by using the semicolon key (;) followed by the comment.

Start the program with a title block, variables and constants. The compiler directive *org* tells the compiler to address the next variable or instruction to the supplied number. Thus, *count_low* is register 8 because *org 8* precedes it. The register after *count_low* would be 9, and so forth. The *device* directive tells the compiler which chip is being programmed, the type of oscillator used, whether the watchdog timer is on or off, and if the security fuse is to be blown:

```
;*******************************
;
;*  PIC® PK Tester V2.0      *
;*  Larry Duarte            *
;*******************************
;
;
;
;
; Variables and equates
;
strobe       =      ra.0    ;noise output
random       =      ra.1    ;50% duty square wave
reset        =      ra.2    ;reset button
statest      =      ra.3    ;status/test button

led1         =      rb      ;port B all LED's
led2         =      rc      ;port C all LED's
             org    8       ;start address of register variable space
count_low    ds     1       ;low byte of delay counter
count_hi     ds     1       ;high byte of delay counter
statdir      ds     1       ;left = 0, right = 1
stat_low     ds     1       ;status counter low byte
stat_hi      ds     1       ;status counter high byte
led1buf      ds     1       ;buffer for LED display
led2buf      ds     1       ;buffer for LED display
strobecnt    ds     1       ;counter for proper strobe length
counter1     ds     1       ;counter used in test function
counter2     ds     1       ;counter used in test function
bitbuf       ds     1       ;buffer to hold random bit
bitcnt       ds     1       ;not used
randomout    ds     1       ;holds random bit
savebit      ds     1       ;old bit
             org    0       ;reset code origin
;
;
```

```
    ; Device data                    CLOCK RATE 78 us
    ;
                        device  pic16c55,rc_osc,wdt_off,protect_off
    ;
```

Recalling the architecture of the PIC®, subroutines must be placed in the first 256 bytes of a page. After resetting *org* to 0, you can start the code with the program's subroutines. It is good practice to name each subroutine with a meaningful label. Example, *waitnochk* is better that *sub1* as *call waitnochk* is much clearer than *call sub1*:

```
        waitnochk                            ;time delay without
            mov    count_low,#0              ;checking for button push
            mov    count_hi,#150
        nochklp
            djnz   count_low,nochklp
            djnz   count_hi,nochklp
            ret
        ;
        wait                                 ;time delay with checking
            mov    count_low,#0              ;for buttons
            mov    count_hi,#10
        waitlp
            csae   ra,#12                     ;ck for any buttons pushed
            ret
            djnz   count_low,waitlp
            djnz   count_hi,waitlp
            ret
        ;
        test                                 ;test for randomness
            mov    counter1,#2               ;lets do it two times
        test1
            mov    counter2,#240             ;2 x 240 = 480 samples
        test2
            mov    led1,counter2
            mov    led2,counter2
            call   getbit
```

```
        call    statupdate
        .....................
        .....................
```

After the subroutines, start the main program at the beginning of the second half of the first page. *Org* resets the next instruction to location 256. The directive *reset start* tells the compiler to set the start-up vector to the label *start* (address 256). The first lines of almost all programs immediately set the configuration for the I/O ports. Secondly, initialize the values of any variables requiring a value:

```
;
; Reset address
;
        org     256
        reset   start                   ;program start
;
start                                   ;HARDWARE INITIALIZATION
        mov     !ra,#15                 ;set RA, all input
        mov     !rb,#0                  ;set RB, all output
        mov     !rc,#0                  ;set RC, all output
        movb    savebit.0,statest
        mov     counter1,#15
timeloop
        mov     led1,counter1
        mov     led2,counter1
        call    waitnochk
        djnz    counter1,timeloop
        mov     led1,#254               ;top led on
        mov     led1buf,#254            ;VARIABLE INITIALIZATION
        mov     led2,#255               ;all leds off
        mov     led2buf,#255
        mov     statdir,#2              ;left - 0
        mov     stat_low,#0             ;right - 1
        mov     stat_hi,#0
        jb      savebit.0,loop          ;CK FOR TEST BUTTON AT START
        call    test
```

```
loop
        jb      reset,ck2           ;IS RESEST BUTTON DOWN?
        ....................
        ....................
```

A common feature of a typical control program is the main loop. After the initialization of the I/O ports and variables, the program steps through a series of *if this...then do...*, then jumps back to the beginning to endlessly repeat the main program loop. An exception to this is an error routine. Upon detecting an error, a control program often shuts off attached devices (motors, communications, etc.) then proceeds to an error-handling routine. It is good practice to leave at least one line available for an LED or buzzer to communicate the controller status to the outside world. In this case, you have a number of LEDs to work with. However, what if you had to show a number of different error codes but only had one LED? Flashing the code is an excellent solution. Example: error code 3 would be 1/2 second on, 1/2 second off, 1/2 second on, 1/2 second off, 1/2 second on, 2 seconds off:

```
delay
        call    wait            ;SPEED DELAY
        jmp     loop
error                           ;ERROR, endless loop
        mov led1,#0
        mov led2,#0
        call waitnochk
        mov led1,#255
        mov led2,#255
        call waitnochk
        jmp error
;
;end
;
```

CHAPTER 8
RS-232 Terminal

The project in Chapter 7 was a self-contained microcontroller application. With the exception of two buttons, it did not interact with the "real world." However, the majority of embedded controller designs are in this arena: the interface and control (hence the name *microcontroller*) of external physical objects. Examples of these objects are motors, displays, sensors, telephone lines, keyboards, lights, and communications. The projects in this book illustrate practical circuits for control of each of these devices. The RS-232 terminal uses three basic external devices: communications, keyboard, and display.

RS-232 is a communication standard for talking to a computer. Established by the *Electronic Industries Association* (EIA), the standard specifies the hardware interface requirements between the sending and receiving devices. The data is sent in a serial format rather than parallel. This means that if there are eight bits to the word, each bit is sent one at a time on a single wire. Parallel communications would send all eight bits simultaneously on eight separate wires. We will explore the specifications later in this chapter.

Why use RS-232? Because almost all PCs and laptops have an RS-232 port. All major computer languages provide support for communicating through an RS-232 port. By adding RS-232 to any controller application, a PC can gain control of all devices attached to the microcontroller. All feedback, button presses, and sensors can be monitored and controlled by the computer. If you added an RS-232 port to our last project, the random LED movement data could be gathered by a PC and a statistical analysis performed on it. Let's proceed and find out just how difficult it is to add this very useful feature.

Objective

A terminal is a peripheral device for computer systems. While terminals may vary, the universal feature found in average terminals is that each has an input method and an output device. All terminals must communicate with the central computer; they have *no internal processing power* (often referred to as "dumb" terminal). Before *graphical user interfaces* (GUI), the standard output was a video display with 80 characters by 24 lines. The input device was a typical QWERTY keyboard. Millions of these types of terminals were connected to mainframes, minicomputers, and even PCs running multi-user operating systems such as UNIX.

A very common communications protocol for all of these systems is RS-232. Some examples of these terminals are the Televideo 950 and Wyse 50/60. An option found on most terminals is a printer port enabling connection to a local printer.

With the advent of *local-area networks* (LANs) in a GUI environment, the text-only terminals are being replaced by much more versatile and easy-to-use terminals. However, before you give them their last rites and bury them in the plot next to slide rulers, they may have an extended life in certain niche markets. Consider code entry for security systems, time clock "punch," industrial machine operator interfaces, and home automation control. All of these applications have a computer system controlling a large area with simple input required from a number of locations. The cost effective way to do this is with "dumb" terminals and NOT with a full PC at each location.

New terminals cost several hundred dollars. However, by using a microcontroller and a handful of parts, you can design a fully-functional terminal for a tenth of the price for a standard terminal! The terminal you will design here will work in many of the applications listed above. To conserve on cost and space, the LCD will only display 16 characters and the keypad will have only 16 keys. For many jobs, this configuration will work fine. If more display characters are needed, replace the 16-character LCD module with a larger one and modify the microcontroller software. (The hardware interface remains the same.) The best way to get the full alphanumeric character set on the keypad involves changing the software to implement a shift key. A hardware change is also necessary for a shift key.

Keyboard Interface

Keyboards come in many different shapes and sizes. Some use hall effect (magnetic switches) to register a keypress, others use capacitive keypads, but most use simple contact closures. You will be using a keypad that has 16 keys of the contact closer type. Most keyboards align keys into a matrix of columns and rows. This keypad is a 4-column by 4-row (4 x 4) matrix. Appearances can be deceiving; a keyboard with no visible pattern will have the underlying circuit board designed to place the keys in an electrical matrix such, as 8 x 16.

The circuit for the terminal keypad is shown in *Figure 8-1a*. Four columns and four rows require eight I/O lines and supports 16 individual keys. The microcontroller is configured so that the rows are outputs and the columns are inputs. With no key pressed, the input's voltage levels are floating. To avoid erroneous readings, pull-up resisters are connected to each of the input lines. To read the keypad, the microcontroller scans the keys. First, all rows

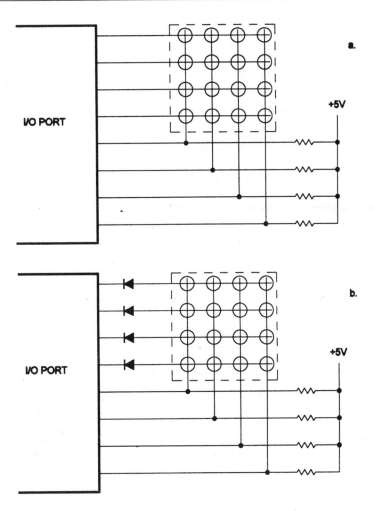

Figure 8-1. A 4 x 4 keyboard matrix yields 16 keys. Add diodes if your device is expected to read more than one key at a time.

(outputs) are set high. Then, one row at a time is set low. If any key on that row is depressed, that key's columns will become low as well. Because the microcontroller already knows which row is low (after all, it made the line low), it simply has to read which column is low to identify the key in the matrix that has been pressed.

This method works great for one keypress at a time (this keypad is designed to work that way), but what if you need to detect two simultaneous keypresses? Let's examine a case where key 1 and key 5 are pressed at the same time. At the start of the scan, all rows are high.

Row 1 is then set low. Since keys 1 and 5 are both pressed, full contact is made between the microcontroller output on row 1 and row 2. This is obviously a problem (and decidedly unhealthy for the microcontroller) since row 1 is being driven low while row 2 is being driven high. The solution is shown in *Figure 8-1b*. By adding diodes to each of the rows, the microcontroller can only sink current and not source it. Now let's reconsider a case where keys 1 and 5 are pressed. Row 1 is set low, bringing column 1 low. The diode on row 2 is reverse-biased, thereby preventing a problem with row 2 driver. These diodes are necessary if you plan to add a shift key.

LCD Interface

LCD stands for liquid-crystal display. It is the perfect choice for small, lightweight, low-power, low-cost text output. LCDs are low powered because they emit no light of their own,

Pin No.	Symbol	Function
1	Vss	Ground
2	Vcc	+5V
3	Vee	Contrast Adjustment
4	RS	Register Select: high for character data low for instruction data
5	R/W	Read/Write: high for read, low for write
6.	E	Enable: enables LCD and clocks in data
7	DB0	Data bit 0
8	DB1	Data bit 1
9	DB2	Data bit 2
10	DB3	Data bit 3
11	DB4	Data bit 4
12	DB5	Data bit 5
13	DB6	Data bit 6
14	DB7	Data bit 7

Table 8-1. Pin descriptions for a liquid crystal display.
This is a very common LCD interface.

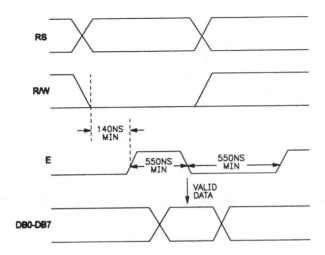

Figure 8-2. Signal timing to complete a write operation to the LCD panel.

but only reflect ambient light. The segments in a LCD can be polarized, permitting light to be passed or blocked. This polarizing current is very low when compared to an LED, which emits light. The display used is a common 16-character by 1-line panel, offered by a number of distributors. The hardware interface is the same for larger units (16 x 2, 16 x 4, 40 x 2), allowing for easy upgrading if you choose. You are not driving the raw LCD segments themselves, but communicating with the LCD controller chip (which itself is a custom micro-controller) which is driving the LCD segments.

The LCD interface consists of eight data lines, three control lines, and three power lines. *Table 8-1* shows the pin assignments for the LCD. A common and useful feature in device specifications is the *timing chart*. *Figure 8-2* is the timing chart for the LCD interface. In order for the device to work, you must not exceed the speed at which the device can respond. This can be done quite easily because of the high MIPS of the microcontroller. Later, we will examine the firmware to see how to put in delays to match the required timing of the LCD.

Let's examine the timing chart in detail. The relevant control and data lines are shown in *Figure 8-2*. A line drawn low or high represents that line at 0V or +5V respectively. When the line is both high and low (as with RS), the line could be high or low at that time. Usually the high or low state represents different options; a high RS means you are sending character data while a low RS means you are sending an instruction. The sequence of events transpires from left to right as follows: first, enable must be low; set read/write to low (write operation), then set RS high (character) or low (instruction). Enable now goes high, but only after a minimum of 140 ns. The data lines (DB0 - DB7) are now set for the character or instruction. Enable is set low, and the data is read into the LCD on the falling edge of enable. The enable pulse must be at least 450 ns. This is the end of the operation; the next operation must wait 1 µs from the previous enable.

In this application, you will never read from the display. Therefore, the R/W line is tied low, representing all write functions. Power for the LCD is +5V and ground. The display contrast is adjusted with a 10K pot between the ground and pin 3 of the LCD interface.

Pin No.	Symbol	Function
1	DCD	Data Carrier Detect: When connected to a modem, this line is active upon detection of a carrier (the frequency data is modulated onto) from another modem.
2	RX	Receive Data: Serial data to the PC.
3	TX	Transmit Data: Serial data from the PC.
4	DTR	Data Terminal Ready: Hardware handshaking, works with DSR to indicate terminal is ready for communications.
5	GND	Signal Ground
6	DSR	Data Set Ready: Hardware handshaking, works with DTR.
7	RTS	Request To Send: Hardware handshaking, works with CTS to limit speed of data being sent. If the input buffer if full then these lines signal a halt to data transmission until buffer is ready.
8	CTS	Clear To Send: Hardware handshaking, works with RTS.
9	RI	Ring Indicator: When connected to a modem, this line is active upon detection of incoming call.

Table 8-2. Interface lines for RS-232 communications. This 9-pin connector is found in PCs, and is referred to as an AT DB9 RS-232 port.

RS-232 Standards

The EIA RS-232 standard is primarily a description of hardware line functions and their respective voltage levels. While many different data communication formats are possible on the hardware RS-232, asynchronous serial has become universal among PCs. The original specification called for a 25-wire interface; however, many lines were unused or had esoteric functions. Today's PCs have a 9-pin serial connector. *Table 8-2* gives the pinouts and functions of each line.

Lines necessary for two-way communications are transmit, receive, and ground. The other lines are used for hardware "handshaking" and can often be ignored. This terminal will used only the three wires required for basic communications. Some PC programs need the handshaking lines in order to get into the proper state before working. A simple trick is to tie CTS to RTS and DTR to DSR on the PC DB9 connector.

Asynchronous serial communications means that you send one bit at a time without regard to any kind of sync or clock pulse. The decoding of the incoming data is perform strictly by timing: you know the start of the data and the fixed baud rate. Baud rate is the speed at which the data is transferred. For example, 9600 baud is 9,600 bits per second. If there are 10 bits to each byte (one start bit, eight data bits, one stop bit), then 9600 baud is 960 bytes per second. *Figure 8-3* shows an example serial byte and timing at 9600 baud. The time between bits (and therefore between samples) is 102 μs. A PIC® operating at 5 MIPS can perform 400 instructions between bits. The firmware codes demonstrate how to translate a parallel byte to serial, serial to parallel byte, and timing between bits.

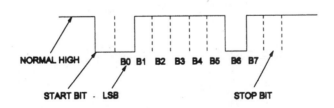

NORMAL HIGH   B0 B1 B2 B3 B4 B5 B6 B7

START BIT - LSB   STOP BIT

9,600 BAUD = 9,600 BITS PER SECOND
ONE BIT = 1/9600 = 104 μS PER BIT
START SAMPLING 52 μS AFTER EDGE OF START BIT

***Figure 8-3**. The character 'A' (ASCII 65) as seen serially with one start bit, one stop bit, and no parity.*

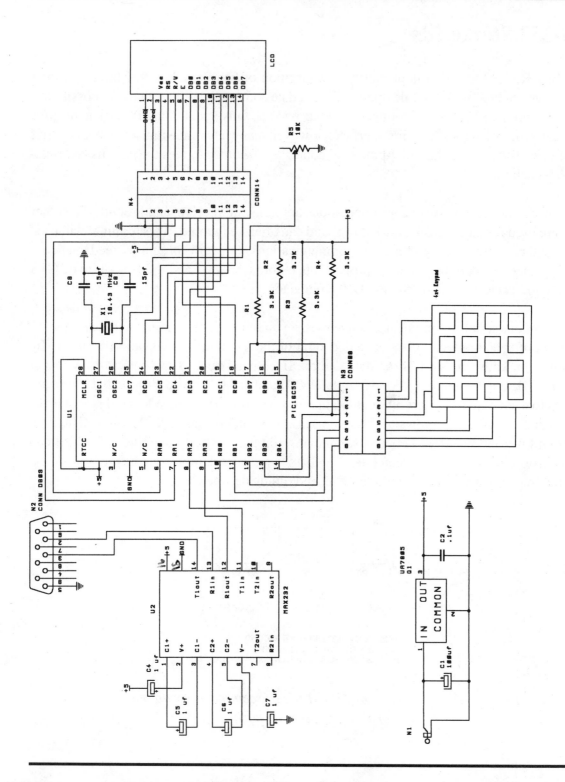

Figure 8-4. A schematic for the RS-232 terminal. The PIC® 16C55 is basically the entire circuit.

Circuit Description

Figure 8-4 is the schematic for the RS-232 terminal. Again, you can see the power of micro-controllers; the PIC® 16C55 is half of the circuit. The receive signal enters through connector N2. It travels to the RS-232 voltage level interface chip, the MAX232. For simplicity, this terminal operates off of a single voltage (an inexpensive wall adapter or 9V battery). However, EIA RS-232 requires +12 and -12 volts. How do you interface the TTL level of the PIC® (+5V) to RS-232 levels? This is where the MAX232 comes in. With just four external *1uf* caps, this chip generates the required voltages and converts RS-232 to TTL. The chip contains four line converters, of which you are using only two for receiving and transmitting. The receive signal leaves the MAX232 and enters the PIC® on pin 9. Internally, the PIC® firmware receives the incoming data and displays it on the LCD, or translates it to special function commands for the LCD. The LCD is byte (8 bits) oriented for data, so you directly connect PIC® port C (8 bits) to the data port of the LCD. This enables a very simple passing of character data by placing incoming bytes onto port C then strobing the LCD control lines.

Transmitting starts at the keyboard. Eight lines are used to read the 4 x 4 matrix, giving you sixteen keys. The PIC® B port is used for this function, the low nibble is configured as output and the high nibble as input. The PIC® firmware scans the keyboard for keypresses. Upon detecting a press, the PIC® translates the key to the programmed ASCII output character (or

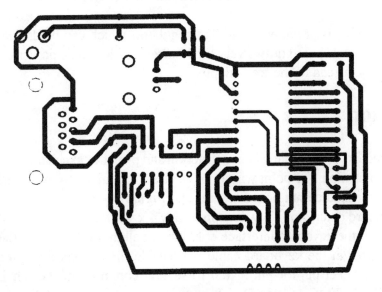

Figure 8-5. The circuit board pattern for the RS-232 terminal.

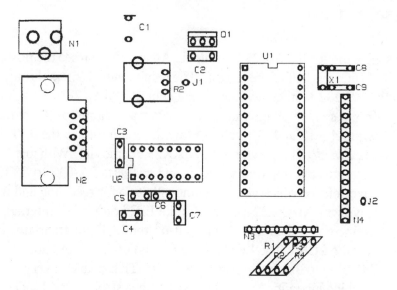

Figure 8-6. If you decide to cut your own board with the pattern in Figure 8-5, here is the part placement layout for the PCB.

string of characters with a simple modification to the program). The serial output is on PIC® pin 8, and is then routed to the MAX232. The MAX232 translates the TTL serial to the RS-232 serial. The transmitted data then leaves the board at connector N2.

Power for the terminal is provided by a 7805 regulator, which can be driven by a DC wall adapter. Since the entire board draws very little current, an alternate power option would be to use a 9V battery instead of the wall adapter.

Construction

This simple project is very forgiving in construction and provides excellent practice for beginners. Wire wrapping, point-to-point wiring, or an etched circuit board will work equally well. *Figure 8-5* is the printed circuit board layout, and *Figure 8-6* shows the part placement for the board. Keyboard connector N3 will support any 4 x 4 keyboard. Consecutive pin-to-pin wiring is possible if the keyboard interface columns are all first followed by the rows. If not, then swapping wires in the keyboard cable will enable the keyboard to work. The LCD connector N4 is a direct pin-for-pin match to several common off-the-self LCD modules, such as the OPTREX DMC-16105. Digi-Key and many other national distributors carry the

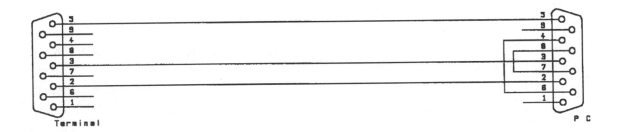

Figure 8-7. *A diagram of a cable between your terminal and an PC serial port.*
Note that the handshaking lines are tied back.

LCDs. Serial connector N2 is a right angle DB9 plug. *Figure 8-7* is a diagram of a cable between the terminal and a PC serial port. Note that the handshaking lines are tied back. *Figure 8-8* shows the completed board for the RS-232 terminal.

Figure 8-8. *The completed board for the RS-232 terminal. The LCD module and keypad is beside the PCB.*

Operation

Connect power to the terminal and the serial cable between the PC and terminal. Most modem programs (Procomm, Comit) have a direct connection feature. Set the communication parameters for 9600 baud, 8 data bits, no parity, and 1 stop bit. Select the direct connection. Now, whatever you type on the PC keyboard will appear on the terminal LCD. Pressing the keyboard Enter will clear the terminal display and send the cursor back to the first character. The terminal also automatically clears the display after 16 characters. Pressing the keyboard will send an ASCII character to the computer; it will appear on the PC screen.

This operation is fundamental and not useful by itself. The real power of the terminal is in conjunction with a custom program which sends prompts to the terminal (*Enter Password*, for example) and expects a limited response from the keyboard (*1234*). Modifications of this elementary terminal will produce dozens of practical applications.

Firmware

The terminal uses an 18.432 MHz crystal, giving you a per-instruction speed of .217 μs. Timing is very important in this program; thus, the need for a calibrated delay routine. This *wait* subroutine can measure time in precise intervals of .651 μs (plus approximately 1 μs of overhead per call). Set the delay desired by loading *count_low* and *count_hi* before the call as follows:

$$delay\ (\mu s) = ((((count\_low - 1) * .651) + .434) * count\_hi) + \\ (((count\_hi - 1) * .651) + .434) + .434$$

You can see that the smallest delay (*count_low* = *count_hi* = *1*) is 1.302 μs, and the longest delay (*count_low* = *count_hi* = *255*, ignoring the case of zero) is 0.0424 second. When *count_low* or *count_hi* is 0, then *count_low-1* or *count_hi-1* is equal to 255 because of the nature of byte subtraction. See Chapter 2:

```
wait                                    ; .217us clock
        djnz count_low,wait             ; full count = .0424 sec
        djnz count_hi,wait
        ret
    ;
```

```
lcd_wait                                ; fix length delay for LCD timing
        clr     count_low               ; delay = 168 us
        mov     count_hi,#1
lcdwait_loop
        djnz    count_low,lcdwait_loop
        djnz    count_hi,lcdwait_loop
        ret
;
wait_check                  ; delay with check for incoming data on comm line
        clr     count_low
        mov     count_hi,#50
ck_loop
        jnb     rs232_in,ck_exit   ;if incoming on rs232, exit now!
        djnz    count_low,ck_loop
        djnz    count_hi,ck_loop
ck_exit
        ret
```

In the subroutine *one_bit*, you send one bit by rotating the output character buffer and placing the carry bit on the transmit line. Then the timer registers are set for a 99 µs delay. This, plus the other instruction times, produces 104 µs between bits (required timing for 9600 baud). A similar procedure is followed for receiving bits in subroutine *get_bit*:

```
one_bit                                 ;send one bit
        rr      out_char                ;set up carry with bit to send
        sc
        clrb    rs232_out
        snc
        setb    rs232_out
        mov     count_low,#152
        mov     count_hi,#1
        ret
;
get_bit                                 ;pull in one bit
        sb      rs232_in                ;test rs232, set up carry for rotate
        clc
        snb     rs232_in
```

```
        stc
        rr     in_char
        mov    count_low,#154
        mov    count_hi,#1
        ret
;
rec_char                         ;get one char from rs232
        mov    count_low,#75
        mov    count_hi,#1
        call   wait
        snb    rs232_in
        ret                      ;false start, get out of here
        mov    temp,#8 ;OK get eight bits
        mov    count_low,#154
        mov    count_hi,#1
rec_loop
        call   wait
        call   get_bit
        djnz   temp,rec_loop
        call   wait
        setb   flag_in           ;rec char ready
        ret
;
send_char
        clrb   rs232_out          ;start bit
        mov    count_low,#158
        mov    count_hi,#1
        call   wait
        mov    temp,#8            ;send eight bits
send_loop
        call   one_bit            ;send single bit
        call   wait
        djnz   temp,send_loop
        setb   rs232_out          ;stop bit
        mov    count_low,#158
```

```
        mov     count_hi,#1
        call    wait
        ret
```

Subroutine *trans_key* translates the keyboard scan code to an ASCII character. Place your desired characters for each key here:

```
trans_key
        csne    key_in,#119
        retw    48              ;return from table with 0 in work register
        csne    key_in,#183
        retw    49
        csne    key_in,#215
        retw    50
        csne    key_in,#231
        retw    51
        csne    key_in,#123
        retw    52
        csne    key_in,#187
        retw    53
        csne    key_in,#219
        retw    54
        csne    key_in,#235
        retw    55
        csne    key_in,#125
        retw    56
        csne    key_in,#189
        retw    57
        csne    key_in,#221
        retw    65
        csne    key_in,#237
        retw    66
        csne    key_in,#126
        retw    67
        csne    key_in,#190
        retw    68
        csne    key_in,#222
```

```
        retw   69
        csne   key_in,#238
        retw   70
        retw   0                    ;default return
```

When communicating with the LCD module, place the LCD delay routing *lcd_wait* between each access to the LCD. This may be overdoing it, but some LCDs are *very* slow:

```
    prt_char
        mov    lcd,char             ;setup data lines for lcd
        call   lcd_wait
        setb   lcd_reg              ;line high for data
        call   lcd_wait
        setb   lcd_clk
        call   lcd_wait
        clrb   lcd_clk              ;strobe clock
        call   lcd_wait
        clrb   lcd_reg
        call   lcd_wait
        inc    char_count
        ret
    prt_loop
        cjne   char_count,#8,prt_char   ;need to change screen?
        mov    lcd,#168             ;change address for part II of screen
        setb   lcd_clk
        call   lcd_wait
        clrb   lcd_clk
        call   lcd_wait
        jmp    prt_char             ;do char now
```

There are two final tricks to remember in this program. When scanning the keyboard, the output scan is on the lower nibble of port B. With no key pressed, the high nibble is held high by the pull-up resisters. *Reading port B (symbol **char**) at this time will always produce 240 or larger* (binary 1111xxxx). Now, consider the case where the scan starts and a key is pressed on the first scan line. The lower nibble is 1110. The low bit pulls down a bit on the high nibble and the total port when read would return 11101110 or 238. This is the scan code and is translated in the previous subroutine *trans_key*. The terminal will scan the keyboard several times during the span of an average keypress. *To prevent sending multiple keystrokes*, save the scan code in *old_key*. Future scans ignore the same scan code until all keys are released clearing *old_key*:

```
        mov     key_in,#255
        mov     keyb,#14
        mov     char,keyb               ;read keyboard
        cjb     char,#240,got_key
        mov     keyb,#13
        mov     char,keyb
        cjb     char,#240,got_key
        mov     keyb,#11
        mov     char,keyb
        cjb     char,#240,got_key
        mov     keyb,#7
        mov     char,keyb
        cjb     char,#240,got_key
        clr     old_key                 ;no key pressed
        jmp     loop
got_key
        cje     char,old_key,noneed     ;same key pressed
        mov     old_key,char            ;save old scan code
        mov     key_in,char
        call    trans_key               ;translate key
        mov     key_in,w
noneed
        jmp     loop
;
;end
;
```

CHAPTER 9
AC Motor Control

Controlling AC line voltages has become a booming application for microcontrollers. Among the many uses is the ability to automatically turn on/off appliances, lights, and motors. Home automation utilizing the X-10 power line communication interface relies heavily on the combination of microcontroller and solid-state switches to accomplish most tasks. The project in this chapter will teach the you basics of AC power line control, as you interface a microcontroller to a motor and a light.

WARNING!

The AC line voltages present in the outlets in your home are extremely dangerous!! Serious injury or death caused by electrocution can occur if you do not follow proper precautions. Do not attempt this project if you have never worked with AC before!

Never touch any part plugged into a wall socket!! Always use a three-wire AC line with a proper wall ground, and attach the ground wire to the case of your project.

While you will be building a fully-functional AC motor controller, this project has no specific function in mind. The motor could be attached to a conveyer belt, garage door, dumbwaiter, or any other application requiring the precise automatic movement of a load. A question may arise as to why an AC motor instead of DC. Besides the desire to explore AC line control in this chapter, an AC motor is much cheaper for the power it delivers and for the simplicity of the required motor power supply (just the AC line!). Using the parts recommended for this controller and the proper AC motor with the gear box, you could lift as much as a 600-pound load.

Objective

The objective is to build a solid-state motor controller that can operate a 1/2-horsepower AC motor. Your design must be able to reverse the motor, provide an optional switched AC load, and have an automatic motor brake. Motor movement (and the attached linear cable) can be measured to a resolution of a tenth of an inch. Maximum travel distance can be set internal to the controller, or operate with limit switches, or both. All external switches are low voltage and can be placed up to one hundred feet from the controller. Finally, since the controller has line AC on it, the board will have a self-contained power supply for the logic circuit (instead of an AC wall adapter).

Stated is the specification of 1/10-inch linear movement resolution. This is really quite arbitrary, and can be more or less depending on the number of notches in a slot wheel (the sensor on the axle of the cable drum) and the diameter of the cable drum. The formula to give you the exact resolution would be: (drum diameter x π)/(number of notches x 2). Thus, your project has a slot wheel with 50 notches and a drum diameter of 3 inches, giving you 0.094-inch resolution. Any of these parameters can be changed within reason.

Three *momentary on* push buttons will operate the controller. In this example, they will be labelled UP, DOWN, and STOP. (Equally suitable would be LEFT, RIGHT, and STOP.) Two limit switches are designed in TOP or LEFT, BOTTOM, or RIGHT, and may or may not be used. The controller is quite capable of running between two points for a very long time without requiring calibration. However, this design does not contain nonvolatile memory, and should there be a power failure, the controller will no longer know its position. Some sort of reset would be required. Even without the power failure, a motor running back and forth a number of times between two points will have a tendency to drift in one direction. Using one limit switch in the UP position as a "home" switch takes care of this problem nicely.

AC Power Control

The fundamental solid-state component behind AC line control is the *triac*. A triac is two back-to-back *silicon control rectifiers* (SCR) in one package. When current is applied to the gate of the SCR, the SCR becomes conductive and passes voltage in one direction until power is removed from the device. Because there are two SCRs in a triac, voltage will pass in both directions once current is applied to the gate. The triac remains turned on until the zero crossing of the AC voltage. At this time, if the gate is not biased, the triac turns completely off. *Figure 9-1* shows the symbol for the SCR and triac, as well as the basic use of a triac in a circuit.

Triacs are very sensitive and require only a few mA of gate current (I_{GT}) to turn on. (This can control 40 amps or more of AC!) This high sensitivity can create problems due to noise and

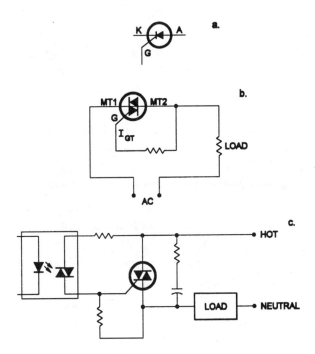

Figure 9-1. *Figure "a" is the symbol for an SCR. "b" shows a triac switching the AC to a load. Controlling the gate current to the triac I_{GT} switches it on or off. "c" is a diagram for an optically coupled AC switch with a snubber network.*

back EMI, especially with inductive (motor) loads. The defining specification for both triac and optocoupler (see the next paragraph) is *dv/dt*, where *dv* is the change in voltage divided by *dt*, the change in time. To eliminate this problem, which causes erroneous firing of the triac, a snubber network is added to the circuit. *Figure 9-1c* is the schematic for a typical triac and snubber network. Teccor designed a special triac for highly inductive loads called an *alternistor*. Even with the alternistor, there can be spurious firing of the triac with disastrous results. To play it safe, both an alternistor and a snubber network for motor control applications are recommended.

You need a method of firing the triac while maintaining low level logic compatibility, and electrical isolation from the logic chips and user interface. An idea solution is the optocoupler. Within the optocoupler is an LED and light-sensitive switch (in this case, another triac). The low level logic turns on the LED which turns on the output switch by *light*! With this combination there is absolutely no conductive path for AC to reach the logic circuity. Let's look at the features of a typical optocoupler, the Siemens IL410. The IL410 guarantees isolation up to 7,500 volts AC. The current required to turn on optocoupler is only 5ma. Somewhat unusual for an optocoupler is the extremely high rating of 10,000 V/μs for *dv/dt*. This part is very immune to transients; the average *dv/dt* is below 1,000 V/μs. You can see the optocoupler in *Figure 9-1c*.

Some optocouplers incorporate a zero crossing circuit. This means that if the signal to turn on the triac arrives when the 60-cycle AC line is above or below a few volts, the optocoupler waits until line voltage is back to zero volts before firing the triac. As mentioned earlier, the triac itself will remain turned on until the line voltage passes through the next zero crossing. Turning on and off at 0 volts is desirable for some applications because it reduces EMI and stress on other electrical components. However, there are definite applications where you *do not* want a zero crossing optocoupler, such as a light dimmer where you need to chop each cycle of the AC line. Zero crossing is of less use in motor applications because of the inductive nature of the load limiting the voltage rise time.

Controller Power Supply

In the previous projects in this book you used an AC wall adapter or battery to provide power for the microcontroller board. Another common design is the incorporation of an on-board 5-volt DC power supply. It makes sense in this particular application because AC is already on the board and can be connected to easily. *Figure 9-2a* shows a typical linear 5Vdc power supply. The transformer provides electrical isolation and steps the 110 volts AC down to 12Vac. The 12Vac is converted to DC by the full bridge rectifier and then filtered by C1. Five-volt regulation is provided by the 7805 and its output filter, C2. The transformer can be selected for 220Vac, or both 110 and 200 by connecting different windings.

Sometimes it is necessary to generate different voltages. If 12Vdc is required, simply change to a higher output voltage on the transformer (such as 16 Vac) and add a 7812 voltage regulator across cap C1. It is a little more difficult to provide negative voltages—a sample power supply is shown in *Figure 9-2b*. The primary difference is a transformer with a split output. Remember, always calculate the current draw and select components sufficient to meet the peak current requirements.

While linear power supplies are cheap and easy, switching power supplies are coming into vogue. They have a much higher efficiency rating and do not require a large, bulky primary transformer. The downside is that they are more expensive and complicated to build. We will

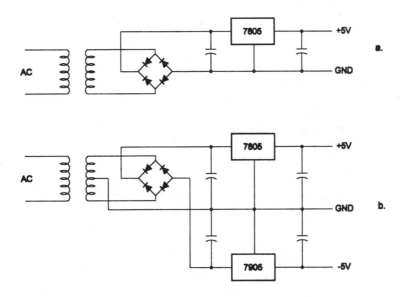

*Figure 9-2. Here is a simple 5-volt linear (a) power supply,
and a dual +5, -5-volt power supply (b).*

not explore switching power supplies, but be aware they are out there, and if your project requires high current, consider a switching power supply.

Circuit Description

The schematic for the AC motor control is shown in *Figure 9-3*. The power supply consists of T1, D1, Q1, C1, and C2. 5Vdc is generated, as discussed above. We use five optocoupler-triac AC switches with snubbers. The AC motor has one line common, plus one line for forward, and one line for reverse. Triac TR1 switches the motor common (line common to both forward and reverse), as well as one side of the motor brake. The motor brake is powered on when the motor is moving, releasing the brake clutch. Triac TR2 switches the other side of the brake. Thus, when the motor is moving in either direction, both TR1 and TR2 are energized, releasing the brake. Removing AC from the motor de-energizes the brake, engaging the brake clutch and stopping the motor. Depending on the direction, either TR3 or TR4 is on at the same time that TR1 is on. Always allow a few seconds between turning on TR3 and TR4 to reverse the motor, because immediate switching between TR3 and TR4 will create back current and overload the triacs. TR5 is for the optional AC load.

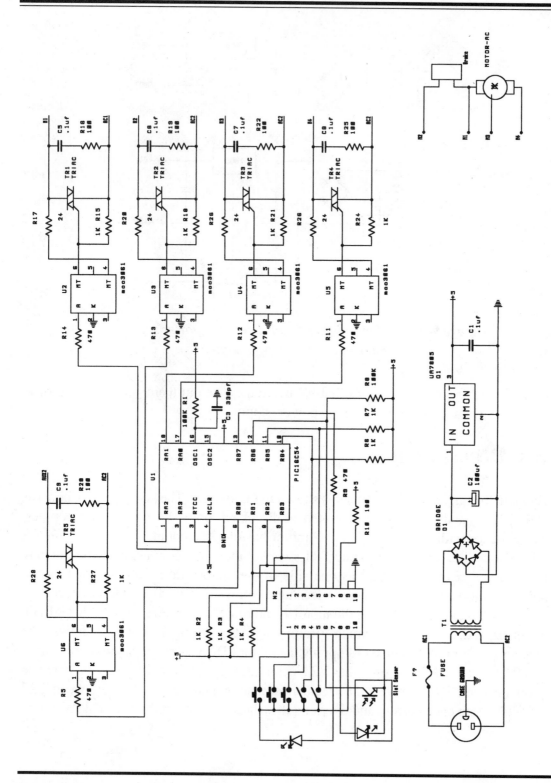

Figure 9-3. A schematic for an AC motor control.

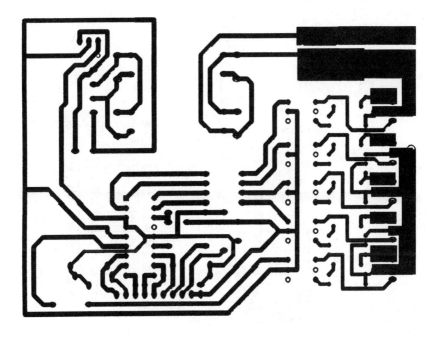

Figure 9-4. *A board layout for an AC motor control.*
Notice the "gap" between high voltage and low voltage traces.

As usual, the microcontroller does all the work. The PIC® 16C54 is connected to three push buttons, two limit switches, and the slot sensor through connector N2. All of these lines have pull-up resisters and are pulled to ground by their respective switch or sensor. Depending on the inputs, the PIC® controls the triacs to perform different functions. The PIC® turns on the triac through a current-limiting resister of 470 ohms to the optocoupler. 470 ohms gives you 10 mA of current, suitable for both the PIC® and optocoupler.

Construction

Figure 9-4 is the parts layout for the board, and *Figure 9-5* is the part placement diagram. Transformer T1 has two sets of holes, allowing for two types of transformers common to the market. Wiring to the motor must be made with 16-AWG wire. Avoid using smaller gauge wire because it will limit the current and could damage the motor. The jumper from J1 to J2 is also 16 AWG. The triacs are aligned on one side of the board, allowing them to be connected to a heatsink, if required. How do you know if you need a heatsink? If your application runs for more than two minutes out of ten, or ever exceeds 20% of the triac's rated

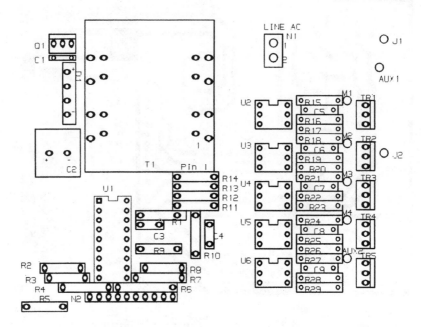

Figure 9-5. *The part placement diagram for the PCB layout in Figure 9-4.*

maximum current, then you most likely need a heatsink. Triacs in a TO-220 package have a metal tab on top which can be connected to line AC. *FOR SAFETY REASONS, ALWAYS PURCHASE ISOLATED TAB TO-220 TRIACS*.   Never heatsink non-isolated tabs together. *Figure 9-6* shows the completed board for the AC motor control.

Line voltage is connected to N1. The ground wire of the AC cord is connected to the case for the board. While the board does not have a fuse on the AC line, one should be used. Either insert the fuse on the case (panel mount fuse), or use an in-line fuse on the cord itself. Wires from N2 to the switches and sensor carry low voltage and can be 22 AWG. Grounds for all the switches are common, and come off of pin 9 of N2. The ground for the slot sensor is pin 10 of N2, and the ground for the status LED can either be pin 9 or 10, depending on your placement of the LED.

A simple notched wheel is used for measuring the distance traveled. The number of notches depends on your desired resolution and the diameter of the cable drum, as discussed previously. Attach the notched wheel to the axle of the cable drum and place the slot sensor to read the wheel, as shown in *Figure 9-7*. The limit switches are placed at the beginning and

Figure 9-6. The completed board for the AC motor control.

end of travel of whatever load is being moved. As mentioned before, these switches are not always used; but it is always a good idea to use at least one switch for the "home" position.

Operation

We have been discussing a general purpose motor controller, and of course, step-by-step operation is dependent on the exact configuration of the final project. To discuss the operation, we will use the possible application of a dumbwaiter lift. This lift travels a distance of ten feet, locks doors to the lift while the lift is moving, and uses one limit switch for the top position.

After construction, power is initially applied to the lift. It does not know where it is, and must be sent to the top position. This is done manually or it can be added to the firmware code to be executed automatically if the lift ever gets "lost." Once the lift is at the top position, the controller stops the lift and resets its internal counters. Pressing the *down* button will activate

Figure 9-7. *The notched wheel is attached to the axle of the motor. The slot sensor sends the motor movement information back to the microcontroller.*

the door locks (consisting of AC solenoids) and send the lift to the bottom position (a distance set in internal firmware). If a mistake is made, pressing *stop* will stop the lift, or pressing *up* will send the lift back to the top position. When the lift is at the bottom, pressing *up* will send the lift to the top position. Whenever the lift is at the top or the bottom position, the doors will unlock.

Firmware

The first highlight of the firmware is in the *device data* line code segment. This project is the only one in the book which has the watchdog timer *on*. In critical microcontroller applications, the watchdog timer is absolutely necessary. Transients, EMI, and even cosmic rays will eventually cause any microcontroller or microprocessor to fail. The watchdog timer will reset the microcontroller, hopefully before any permanent damage occurs. The price to pay for this protection is making sure that the statement *clr wdt* is executed periodically. The kind of pitfall to look for is in the *wait* subroutine. Upon power-up, the default time-out period for the watchdog timer is 2.5 seconds (the maximum). The *wait* loop can last 30 seconds. Obviously, with no *clr wdt* instruction in the loop, the program will reset every time *wait* is called with a time over 2.5 seconds. All calls to *wait* in this program are under 1 second, but be careful in your own programs:

```
;
; Device data
;
                    device  pic16c54,rc_osc,wdt_on,protect_off

;

wait                                ; 38us clock, 152us per instruction
        djnz    count_low,wait      ; 3 cycles + 3 more every 256
        djnz    count_hi,wait       ; full count = 30 seconds
        ret
```

The *start_down* routine is typical of the other routines in this program, and shows the intelligence that a microcontroller adds to this simple application. Upon starting down, you first check to see if the bottom limit switch is tripped. If so, don't go down. Then, check to see if you are at or below the "set" distance. If so, don't go down. Then, is the lift currently moving? If yes, and it's moving in the down direction, ignore the command. If yes, and the lift is moving up, stop the lift, wait one second, and reverse its direction. Notice the *clr wdt* instruction after the call to *wait*. If the lift is not moving, then immediately start the lift down:

```
start_down
        jnb     limit_bottom,down_exit          ;at bottom limit, return
        cjb     drum_count_hi,#set_hi,down_ok   ;chk set limit
        cja     drum_count_hi,#set_hi,down_exit    ;way too low, return
        cjb     drum_count_low,#set_low,down_ok    ;chk set limit
        ret                                     ;all ready at set position
down_ok
        jnb     flag_pause,down_now             ;not moving
        jnb     flag_direction,down_exit        ;all ready moving down?
        call    start_stop                      ;must be moving up, stop it
        clr     count_low
        mov     count_hi,#8                     ;wait 1 seconds
        clr     wdt                             ;clear watchdog timer
        call    wait
down_now
        setb    flag_pause                      ;moving
        clrb    flag_direction                  ;downward
        mov     count2,#stuck_slot*2            ;stuck drum counter
        movb    flag_slot_state,drum            ;save drum state
        mov     ra,#motor_down
down_exit
        ret
;
```

The slot sensor is used for safety as well as position information. If the motor is moving and you don't see a change in state of the slot sensor, then it is in a severe error condition. The "stuck slot" counter is loaded with a predetermined count before the start of motor movement. As you go through the *main_loop*, this counter is decremented. If it reaches 0, then an error has occurred. Only when a change in the slot sensor is detected will the counter be reloaded with the "stuck slot" value. With this method, you can detect a lift problem a fraction of a second after it occurs. In the following *main_loop*, look for the next three functions. The watchdog timer is cleared on every pass. The "stuck slot" counter is decremented or reloaded. A final technique for critical applications is the resetting of the I/O port configuration. A glitch may reset the configured lines as *in* or *out*. The program itself will continue to run, so the watchdog timer will never reset the program; but you lose control of the I/O lines. Periodically resetting the I/O direction is recommended by Microchip:

```
main_loop
        lset    main_loop               ;reset page
        clr     wdt                     ;clear watchdog timer
        mov     !ra,#0                  ;reset IO direction ra, rb
        mov     !rb,#126                ;in case of hiccup
ml_ck_buts
        sb      but_down                ;is down pressed?
        call    start_down              ;process down button
        sb      but_up                  ;is up pressed?
        call    start_up                ;process up button
        sb      but_stop                ;is stop pressed?
        call    start_stop
        jnb     flag_pause,main_loop    ;we are not moving
ml_ck_sensor
        djnz    count2,ml_not_stuck     ;is drum moving?
        mov     ra,#motor_stop          ;STUCK!
        mov     error_code,#4
        jmp     error_end
ml_not_stuck
        jnb     flag_direction,ml_ck_down  ;ck for limits & position - up or down
        jb      limit_top,ml_drum_slot     ;jmp if not up
        call    start_stop              ;we are up
        clr     drum_count_low
        clr     drum_count_hi           ;reset drum counter
        jmp     main_loop               ;stop
ml_ck_down
        jnb     limit_bottom,down_stop
        cjb     drum_count_hi,#set_hi,ml_drum_slot    ;chk set limit
        cjb     drum_count_low,#set_low,ml_drum_slot  ;chk set limit
down_stop
        call    start_stop              ;we are at set position
        jmp     main_loop
ml_drum_slot
        jb      flag_slot_state,ml_low_state   ;look at old state
```

```
        jnb    drum,main_loop              ;is drum low, if so jump
        jmp    ml_do_math
ml_low_state
        jb     drum,main_loop              ;is drum hi, if so jump
ml_do_math
        movb   flag_slot_state,drum        ;save drum state
        mov    count2,#stuck_slot   ;detected movement, reset stuck counter
        jb     flag_direction,ml_up_math   ;add or subtract
        ijnz   drum_count_low,main_loop ;increment and jump if not zero
        inc    drum_count_hi
        jmp    main_loop
ml_up_math
        dec    drum_count_low              ;going up, decrement counter
        cjne   drum_count_low,#255,main_loop
        dec    drum_count_hi
        jmp    main_loop
```

CHAPTER 10
A Telephone Busy Buster

A friend of mine was anxiously awaiting Monday morning to arrive, because at 10:00 AM, Garth Brooks tickets were going on sale and she greatly desired going to the concert. Upon the appointed hour she began calling Ticket Master to purchase the tickets. The line was busy. She called again, and again the line was busy. Between cups of coffee, housework, and a little TV she kept calling. The same torturous response greeted her each time, the plaintive beeping of a busy signal. Finally, nearly two hours later, she got through, but the concert was sold out.

Another friend had a similar problem. He was an avid golfer, a member of several private golf clubs. Like most players, the weekends were the best time for him to play. Golf courses often take reservations for weekend tee times on a certain day, starting at a certain time (typically Wednesday, at 7:00 AM) in order to give everyone a fair shot. As a result, each week he wasted hours on the phone trying to get through to get a reservation. (It seems the competition to get a good tee time is greater than the struggle to lower your handicap).

Have you ever tried reaching a lottery-winning number line after the night of the big drawing? You have a better chance winning the "big one." How about calling talk radio shows (I've tried for hours to get through), or "call now for the thousand dollar prizes"? Government agencies (DMV, IRS) often have busy lines as well.

It's clear we need yet another electronic gadget to help make our lives easier and more productive—we need the Busy Buster. Some phones have a redial button on them, but this obviously requires you to waste time waiting by the phone, pressing the redial button time and time again. A few models have an automatic redial feature, but the timing is fixed at a slow redial rate (every ten minutes) and shuts off after one hour. The Busy Buster, on the other hand, is a demon dialer (every five seconds!), and has the dogged determination of a pit-bull terrier.

Objective

What is needed is a device that is compatible with the average household touch-tone phone. It should have a modular input and output jack, enabling it to be placed between your phone and the wall outlet. The device should be small and unobtrusive. The ideal power supply would be power from the phone line, but unfortunately, the current draw of the Busy Buster is just over what the phone line can provide. It is low enough, however, to run for a long time off of a 9-volt battery or a small AC wall adapter. A local phone call consists of a seven-digit number, and a long-distance number (or toll-free, 1-800 number) is eleven digits. While unusual, there may be times when you will need the Busy Buster for long-distance calls, so it therefore should be able to store eleven digits.

What about the user interface? As always, simplicity is the goal. How should the phone number be entered into the Busy Buster for redialing? A keypad on the device is redundant because there is already one close at hand: the device should 'read' the phone number directly from the phone! Once the number is captured, redialing commences and continues until the device no longer receives a busy signal.

Upon normal ringing, the user must be alerted to pick up the phone. An LED indicator is no good as it would force you to constantly watch the device. The solution is a loud beeping sound when the phone is ready to be picked up. This will enable you to do other chores while waiting for the phone. Thus, the entire operation is reduced to turning on the Busy Buster, dialing a number, hanging up the phone, and waiting for the audible indicator. It just doesn't get any easier.

Telephone Line Standards

The telephone system has been in operation for around 100 years. Surprisingly little has changed with the basic workings of your home phone. There are many books available on the technology behind today's phone system; we will study just enough to accomplish the stated goals. The typical home telephone junction box contains four wires; green (tip), red (ring), yellow and black. The average home phone uses only tip and ring. The phone company provides voltage on the line which is modified by the diaphragm of the handset. This varying voltage is connected to the receiver by a network of wires, switches, microwave and fiber-optics. The speaker in the telephone handset receives the varying voltage and recreates the exact sound from the original source. *Figure 10-1* is a rough representation of this process.

While this project does not deal with the vocal aspect of the telephone connection, you now know that there are only two wires to work with, tip and ring. How do you initiate a call, provide the phone company with the desired number, and receive call status (ringing, busy) signals? When a phone is hung up (on the hook), the tip/ring is open. Lifting the phone (off

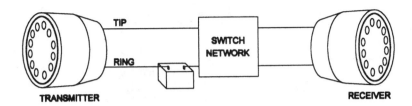

Figure 10-1. *In a phone connection, the diaphragm of the transmitting handset varies the voltage to the receiver, which in turn recreates the original sound.*

the hook) places a load on the tip/ring, reducing the nominal 48 volts to about 5 volts. The central office can now receive the phone number. There are two common methods of transmitting the desired phone number: pulse and touch-tone. Your device will only work with touch-tone. This is not a problem since most modern phones (and telephone switching systems) support touch-tone dialing. The phone transmits the number by using DTMF (dual-tone multifrequency). The central office, via switching networks, will connect you to the dialed number. Upon reaching the destination you will receive status signals. By varying the cadence and frequency of the signal, you know if the phone is ringing, busy, or otherwise.

The phone system is a public utility under the control of the Federal Communications Commission (FCC). Any device connected to public phone lines must conform to Part 68 of the FCC rules. Among the many rules are specifications for impedance matching, hazardous voltage protection, hearing aid compatibility, billing protection, signal power, leakage current, and much more. One interesting rule that relates to this project is, *"Note that repetitive dialing to a particular number by all types of automatic dialers must cease after 15 attempts. There are no limitations when calling is made sequentially to two or more alternative numbers, or when spaced 10 minutes to a single number."* In the same FCC pamphlet is the qualification, *"In Docket No. CC-81-216, Fourth Notice of Proposed Rulemaking (FCC 86-352), the Commission determined on August 5, 1986 that Part 68 limitations on computer-controlled automatic dialing were unnecessary at the time of the decision. The Commission said it would revisit the issue if necessary to ensure network protection."* Where does this leave us? Check with the FCC for the most recent rules before attaching any devices to your phone lines.

DTMF Interface

To meet the objective, you need to be able to decode the DTMF signals (so you can store the dialed number in memory), generate DTMF signals, and process phone status signals. This is

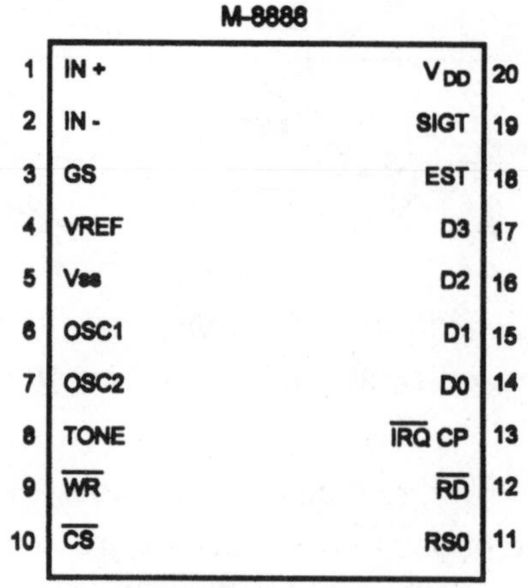

M-8888

```
   1 | IN +              V DD | 20
   2 | IN -              SIGT | 19
   3 | GS                 EST | 18
   4 | VREF               D3  | 17
   5 | Vss                D2  | 16
   6 | OSC1               D1  | 15
   7 | OSC2               D0  | 14
   8 | TONE            IRQ CP | 13
   9 | WR                 RD  | 12
  10 | CS                 RS0 | 11
```

PINS	FUNCTIONS
IN +, IN -	TONE INPUT
GS	GAIN FOR INPUT
VREF	CONNECT TO IN + FOR SINGLE ENDED INPUT
Vss	GROUND
OSC1, OSC2	OSCILLATOR INPUT
TONE	DTMF OUTPUT
WR	LOW WILL WRITE TO M8888
CS	LOW ENABLES CHIP
RS0	REGISTER SELECT ADDRESS BIT
RD	LOW WILL READ DATA FROM M8888
IRQ/CP	INTERRUPT REQUEST OR CALL PROGRESS OUTPUT
D0, D1, D2, D3	DATA BITS
EST, SC/GT	STEERING INPUTS FOR FOR VALID TONE PAIR
V DD	+5V

Figure 10-2. The pinout of the M-8888. This chip can receive and transmit DTMF signals, and provides call progress filters.

accomplished by the Teltone M-8888 chip. Its many features include low power consumption, a complete DTMF transmitter and receiver, a single +5 volt supply, call progress filters and microprocessor port operation. *Figure 10-2* shows the pinout of the M-8888. Each DTMF tone consist of two frequencies. *Table 10-1* gives the frequencies for each digit on the phone keypad and the binary equivalent on the data lines (D0 - D3) of the M-8888. An external 3.58-MHz crystal allows precise decoding and generation of these frequencies.

The M-8888 has several different modes of operation depending on the configuration of its control registers. Control register A will enable DTMF (generate and receive tones), CP (call progress) and IRQ (interrupt enable) modes of operation. DTMF and CP modes are mutually exclusive; therefore, after dialing a number in DTMF mode you would reconfigure the M-8888 to CP mode for detecting busy or ringing signals. If IRQ mode is on, then the M-

1st Freq.	2nd Freq.	Digit	D3	D2	D1	D0
697	1209	1	0	0	0	1
697	1336	2	0	0	1	0
697	1477	3	0	0	1	1
770	1209	4	0	1	0	0
770	1336	5	0	1	0	1
770	1477	6	0	1	1	0
852	1209	7	0	1	1	1
852	1336	8	1	0	0	0
852	1477	9	1	0	0	1
941	1336	0	1	0	1	0
941	1209	*	1	0	1	1
941	1477	#	1	1	0	0
697	1633	A	1	1	0	1
770	1633	B	1	1	1	0
852	1633	C	1	1	1	1
941	1633	D	0	0	0	0

Table 10-1. Dual tone frequencies for phone keys and the equivalent binary coding from the M-8888.

8888 would indicate a valid DTMF tone has been received by pulling the IRQ pin low. Also, if IRQ mode and BURST mode is on, then IRQ pin will go low when the transmitter is done with the current tone. Control register B contains several test functions and the flag for BURST mode operation. In BURST mode the transmitter will output a fixed-length tone followed by a fixed-length pause. This is good for high-speed dialing.

The microprocessor port provides easy interfacing to the microcontroller. Pin RS0 selects between the transmit/receive buffer (low) and the control registers (high). A low on RD will read the data, while a low on WR will write data to the register or buffer. During any reading or writing to the M-8888, CS (chip select) must be low. Pins D0-D3 are bidirectional data

Progress Tone	1st Freq.	2nd Freq.	On Time	Off Time
Dial	350	440	Continuous	
Busy	480	620	0.5	0.5
Ringback, Normal	440	480	2	4
Ringback, PBX	440	480	1	3
Congestion	480	620	0.2	0.3
Reorder	480	620	0.3	0.2
Receiver Off-hook	1400 + 2060	2450 + 2600	0.1	0.1
No Such Number	200 to 400		Continuous	

Table 10-2. *Call progress tones from your phone.*
Both frequencies are added for a dual-tone.

lines. This means that when reading from the chip, the data lines are outputs, and when writing to the chip, the data lines are inputs. The IRQ/CP pin (open-drain output, requires pull-up) serves two functions. In DTMF mode this pin is low it's when ready to transmit the next tone, or when you have received a valid tone on the input. In CP mode the M-8888 toggles this line upon receipt of a valid call progress frequency.

Detecting call progress signals deserves further explanation. *Table 10-2* lists common frequency and cadence for progress signals in the United States. The M-8888 call progress filter converts the raw tone frequency into a square wave matching the cadence of the signal. Thus, the microcontroller analyzes the timing of the cadence to determine the status of the line. *Figure 10-3* shows us this conversion process.

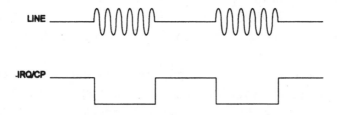

Figure 10-3. *Call progress tones are converted into highs and lows by the M-8888.*

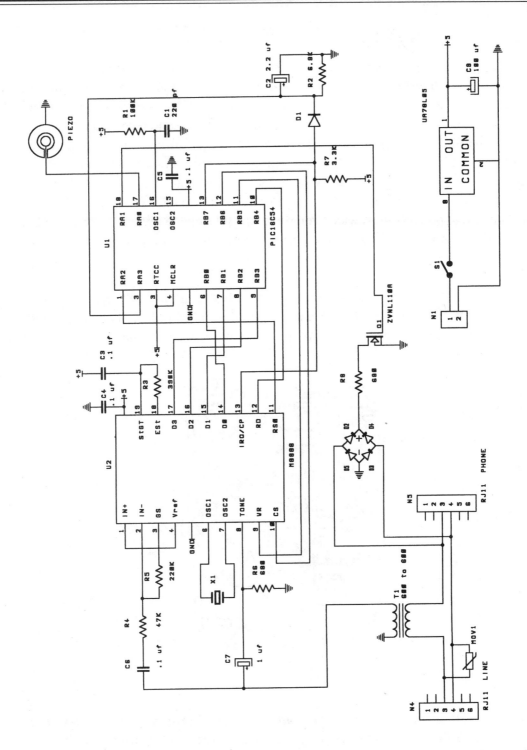

Figure 10-4. The Busy Buster schematic. *Two chips do it all.*

Circuit Description

The schematic for the Busy Buster is shown in *Figure 10-4*. Let's consider the interface to the phone line. The incoming line is connected to N4. MOV1 is across the tip/ring for spike protection. T1 is a 600-to-600 ohm telephone transformer used to receive and inject DTMF tones into the line. The tip/ring is then connected to the full-wave rectifier consisting of diodes D2, D3, D4 and D5. This bridge is not absolutely necessary, but is a useful safety precaution. Since the voltage on the tip/ring is DC, it is possible (caused by bad line wiring in the home, or by incorrect Busy Buster construction, etc.) to have incorrect polarity on the lines. The bridge protects the device even if the tip and ring are reversed. R8 is the load for the line. Q1 is an n-channel MOSFET and provides switching to turn the load (and the phone line) on or off.

Both receive and transmit are capacitively coupled between the M-8888 and T1. The M-8888 DTMF input amplifier can be either differential or single-ended. By tying pin 1 to pin 4 we have chosen the single-ended configuration. R4 and R5 sets the gain for the input. C6 is

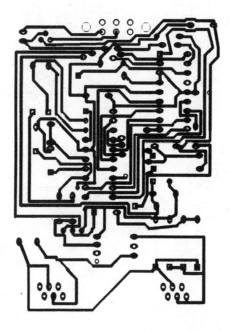

Figure 10-5. *Busy Buster PCB layout. To keep the size small, the traces are relatively tight. Be careful if you decide to etch your own board.*

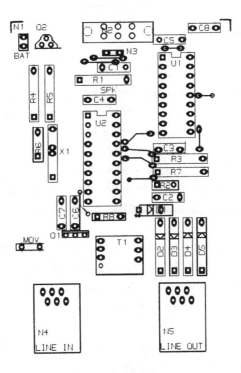

Figure 10-6. The part placements for the layout in Figure 10-5.
Don't forget the jumpers.

the input capacitor. R6 and C7 provide the DTMF output to T1. R3 and C3 provides the RC timing constant for the 'steering circuit.' Basically, this is the time which a tone must appear at the input to be registered as a 'good' input. R7 is the required pull-up for IRQ/CP pin. D1, C2, and R2 comprise a further filtering network for call progress signals.

Power for the board is regulated by Q2, a 78L05. The input source in this schematic is a 9-volt battery, but it could just as easily be an AC wall adapter with a 9-volt DC output. The speaker is a piezo element type and can be driven directly from a TTL logic level.

The brains of the Busy Buster is the PIC® 16C54. While this PIC® is Microchip's smallest and least powerful microcontroller, it can do the job beautifully with room to spare. In fact, only 35% of its program memory is used!. The firmware is coded for an 11-digit phone number and the PIC® has enough RAM to add ten more digits if required. R1 and C1 provides the RC constant for the PIC's® clock. The raw clock frequency is divided by four to give you a per-instruction time of 108us. I/O line RA0 drives the peizo speaker. RA1 controls the on-off hook switch. All other I/O lines are attached to the M-8888.

Construction

Figure 10-6 is the parts placement diagram, and *Figure 10-5* shows the PCB layout. As you can see, in order to meet the goal of a small compact unit, the components on this board are fairly close together. Care must be taken to prevent shorts during soldering, plus you need to be careful of the traces bleeding together if you cut your own board. The modular phone jacks are held in place by the six pins toward the rear; the front of the jack should be glued to the PCB for extra strength. Power connector N1 may be wired to a 9-volt battery cap or to a panel mount power jack. Wires from the peizo speaker are directly soldered to connector N3. *Figure 10-7* is the completed board for the Busy Buster.

If you choose to put the unit in a case, choose a box the same length as the board. Cut holes in the top for the power switch, in the bottom for the phone plugs, and in the front for the speaker. When connecting the Busy Buster to the line, be sure N4 is connected to the line wall plug, and that N5 is connected to your phone.

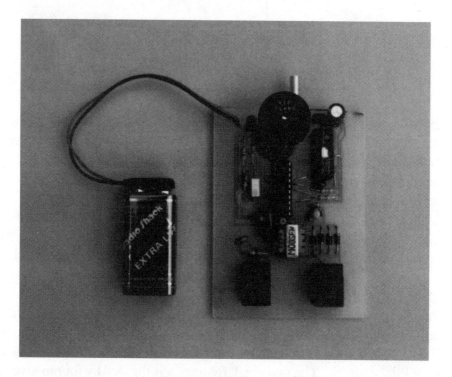

Figure 10-7. *The completed board for the Busy Buster. This one has a battery attached to make it portable—it could have had an AC adapter for a permanent installation.*

Operation

If you know ahead of time that a busy signal is likely, then turn on the Busy Buster. Alternatively, once receiving the busy signal, hang up and flip on the power switch. When powered on, the Busy Buster will beep twice to let you know it's ready. Dial the number. If no busy signal is received, turn off the unit and proceed with the phone call. If you receive a busy signal, touch the # key on the phone. This tells the unit to start redialing. Hang up the phone after pressing #. The Busy Buster emits one short beep each time it dials. This way you know it's doing its job, and you can see the speed at which it is operating. Upon receiving a ringing status instead of a busy signal, the unit will beep continuously—the Busy Buster is holding the line open for you. Pick up the phone and turn off the device. Sometimes, when a line receives a lot of traffic, the central phone office will divert the call to an error message. This message is preceded with a ringing status. Obviously, the Busy Buster cannot tell this apart from a 'normal' ring and will start beeping. As in the case of making an error in dialing, reset the unit and redial.

I tested the unit on a convenient number, the local lottery winning number line. The average time to get through without the Busy Buster was five minutes while dialing every 15 seconds or so. With the Busy Buster in use, the average time to get through was 25 seconds! On this local number the unit could complete a cycle (dial, determine status, hang up) in five seconds. At the beginning of this article, I listed several cases in which this device would come in handy. After testing the Busy Buster, it is now evident that not only will this free your time from constantly dialing, the rapid rate of redialing will increase your chances of getting through quickly on a busy line.

Firmware

In the firmware for this project you will see a good example of indirect addressing. Direct addressing is what is used most of the time; its format is *mov test,#24*. This command moves the number 24 to register *test*, which was previously defined as register number 20. But what if you want to perform an identical set of instructions on a number of registers? One solution is to rewrite those instruction over and over, each time inserting the new register number. A much better solution is to write the instructions once within a loop using indirect addressing to reference each register per loop. Let's see how this works. In the opening definitions, set aside a block of consecutive memory for the phone number:

```
pn1     ds      1       ;phone number - register 12
pn2     ds      1
pn3     ds      1
pn4     ds      1
pn5     ds      1
pn6     ds      1
pn7     ds      1
pn8     ds      1
pn9     ds      1
pn10    ds      1
pn1     ds      1       ;phone number - register 22
```

The phone number will now be stored in RAM locations 12 to 22. The FSR (file select register) is used to point to each register of the phone number. Let's load FSR with the number 12. Remember the previous example of direct addressing and compare it to *mov indirect,#24*. The indirect keyword tells the compiler to use indirect addressing in the *mov* command. The number 24 will be moved to the register that FSR is pointing to, in this case register 12. By simply incrementing FSR, you are now ready to address the next byte in the phone number, register 13. Look at the subroutine *dial*, and you will see this exact method of addressing the phone number:

```
dial
        mov     fsr, #12                ;prepare FSR for indirect addressing
dial2
        mov     data,indirect
        call    w8888
        inc     fsr                     ;increment FSR
        csb     fsr,pnlen               ;compare to phone number length
        jmp     dialdone
        mov     counthi,#2              ;164ms
        mov     count_low,#250
        call    wait
        jmp     dial2
dialdone
        setb    rso
```

```
        call    r8888               ;clr interupt
        clrb    rso
        ret
```

In the main loop there is a need to use the PIC's® RTCC (real time clock/counter). By keeping track of the pulse length of the call progress signal, you can determine the cadence and therefore the current phone line status signal. When the CP line goes high, clear the RTCC. You will then go into a tight loop looking for the CP line to drop low. When this happens, you can read the RTCC. Depending on its value (which represents the time), you have a busy signal, ringing, or an unknown condition. If you have an unknown condition, you try again:

```
        call    dial
cklowset
        mov     temp1,ra            ;make sure det is low to start
        and     temp1,#1000b
        csb     temp1,#1
        jmp     cklowset            ;its high so goback
detlow
        mov     temp1,ra
        and     temp1,#1000b        ;only det is left
        csb     temp1,#1            ;checking for high or low on det
        jmp     dethigh             ;det is high
        jmp     detlow
dethigh
        mov     rtcc,#0
dethigh2
        mov     temp1,ra
        and     temp1,#1000b        ;only det is left
        csb     temp1,#1            ;checking for high or low on det
        jmp     dethigh2            ;det is high
        mov     temp1,rtcc          ;det is low, look at time
        csb     temp1,#14           ;less than .4 sec?
        jmp     cktme               ;no
        jmp     detlow              ;back to start
```

```
cktme
        csb    temp1,#25        ;less than .7 sec
        jmp    aok              ;no, ringing
        jmp    mkcall           ;we got busy
aok
        call   beep             ;endless beeps
        jmp    aok
```

CHAPTER 11
A Video Output Thermometer

How cold is it? Is it time to turn up the thermostat or to put on a sweater? Are the kids properly dressed for school? You could used that old dial thermometer on the porch or the mercury thermometer on the window sill to find out, but they are hard to read and usually inaccurate. What you need is a convenient, easy-to-read, indoor/outdoor thermometer. The solution is a video thermometer. With the flick of a switch, you will be able to see both the inside and outdoor temperature, accurate to within one degree!

As an added bonus to the functionality of this project, you will be pushing a standard off-the-shelf microcontroller (PIC® 16C71) to the limits of its capabilities. This is a general purpose microcontroller, never intended to generate video. With just a few discrete parts and the proper firmware you will be able to produce an acceptable NTSC compatible signal. When you generate the constant video signal, will you have enough power to read not one but *two* temperature sensors and format the changing numbers into information for the display? Yes. In fact, you will not even use all the PIC® RAM, plus you will use only 75% of the available code space. The unit in operation is shown in *Figure 11-1*.

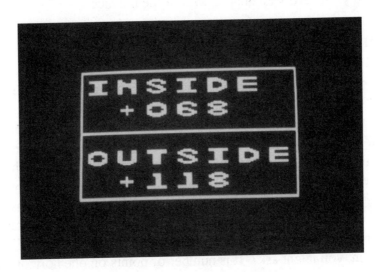

Figure 11-1. *The output from the video thermometer.*
Both inside and outside temperatures can be displayed on a standard TV.

Objective

The first requirement is to accurately sense the temperature. You should be able to read from 50 below zero to over 150 degrees Fahrenheit within one degree of accuracy. This range is most likely unnecessary for comfortable living environments, but it makes the device useful for monitoring the temperature inside a refrigerator or an electronic equipment rack. The sensor should also be small and durable to withstand harsh environments. Finally, the temperature sensor should be precalibrated, immune to signal loss due to long cable runs from the microcontroller, and TTL compatible.

The output from the device must be a NTSC video signal. It will drive a 75 ohm video input jack (standard impedance) on an average monitor or TV. While it would also be nice to provide color, the microcontroller is not quite powerful enough to do that. Instead, you will generate a sharp black-and-white signal with the NTSC video, which will work on most modern monitors and TVs.

The microcontroller must have enough I/O lines to read two temperature sensors and produce a video signal with sync, black and white levels. Of course, the controller must also be fast and accurate enough to generate the rigorous timing of the video signal. The video thermometer is expected to be in constant use, so you need a power supply that is longer lasting than a battery. An AC wall adapter will fit the bill nicely.

Video Standard

Okay, let's now get into the nuts and bolts by analyzing the signal that you are trying to produce. A TV produces a picture by varying the intensity of an electron beam scanning across a CRT (cathode ray tube). *Figure 11-2* is a simplistic representation of this process. The video signal must contain all of the information in order to synchronize the CRT and 'paint' the picture. In the United States, video standards were set in the 1940s. These standards are referred to under the NTSC (National Television System Committee), and allowed TVs across North America to receive all commercial broadcast signals. This has also limited us to the technology of the time in terms of resolution. Maximum resolution is based on the bandwidth of the signal, which is limited by the spacing of the RF carrier. What is meant by resolution, you might ask? It is number of points on one horizontal line if it alternated between white and black. When determining the video standard, the NTSC concluded that 700 points per line, 525 horizontal lines 60 times per second, would produce a sharp

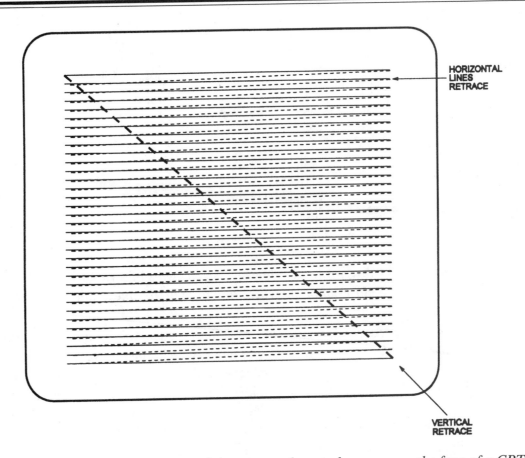

Figure 11-2. *A standard NTSC signal sweeps an electron beam across the face of a CRT. Horizontal and vertical retraces are dotted lines; they are not normally visible.*

flicker-free image. The bandwidth of this signal would be 350 cycles (700 alternating black and white) x 525 lines x 60 per second = 11,025,000 cps. This 11 MHz bandwidth is double the allotted space assigned by the FCC. To handle the problem, broadcasters introduced interlaced scanning. One frame (a complete top-to-bottom scan of the CRT) consists of 262.5 lines; each line is either the odd line or even line of the original 525-line frame. The video signal alternates between the odd and even lines, giving the appearance of 525 lines at half the bandwidth.

In addition to the actual picture information, the video signal must also contain horizontal sync pulses and vertical sync pulses. *Figure 11-3* presents the complete video signal for one 262-line frame. The pulses from 0.25 volts to 0 volts are for the horizontal sync. One second

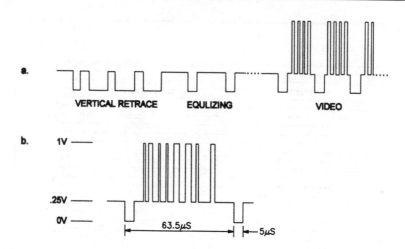

Figure 11-3. *Waveform "a" show all the elements of a video signal.
"b" is a close-up of one video line from a 262-line frame.*

divided by 60 frames per second, divided by 262.5 lines is equal to 63.5 μs. The heartbeat of the video signal is the 63.5 μs horizontal sync pulse, which must be very consistent or an unstable picture will result. The 0.25 voltage level of the signal is called the blanking level, and is black on the CRT. By using a voltage level below black to represent the retrace time, the retrace is cleverly hidden from view. There are 60 frames per second, so the vertical retrace must occur within every 1/60th of a second, or 16.6 ms. The vertical retrace signal is the three 'long' 0-voltage level pulses seen at the beginning of the trace in *Figure 11-3*. Again, this time period is below black and therefore not visible to the viewer. Following the vertical retrace period is equalizing pulses. These pulses are used for shifting the image 1/2 line down for interlace scanning. This project uses non-interlace scanning; therefore, the equalizing pulses are not used.

Finally, the picture information itself resides between the horizontal pulses; one volt for maximum white, and 0.25 volt for black. This full-on approach is generally not used in a broadcast video signal, and may be too intense for some TVs or monitors. You can adjust the voltage level by simply varying one resistor.

Temperature Sensor

The most common temperature sensor is the thermistor. The thermistor is a resistor that decreases in resistance as the temperature increases. By adding this to a resistor network, the output voltage will change with temperature. The deficits to the thermistor are several: it must be calibrated; it is affected by long cable lengths; and an analog-to-digital converter is needed to read the voltage. The next step up is a temperature sensor that National Semiconductor makes, the LM45. It has a fixed linear output (10 mV/C), and requires no external calibration, but you still have to deal with line loss and input to an A/D.

This project uses one of the most sophisticated temperature sensors on the market today, the Dallas Semiconductor DS1620. Some of the many features include: no external components; a temperature range of -55°C to +125°C; temperature sensing every 200 ms; direct microprocessor compatibility; calibration and accuracy to within 0.5°C; and immunity to cable distance (within reason). This chip can operate in several different modes, including a stand-alone mode that triggers a load when a certain temperature is exceeded. *Figure 11-4* shows the pin assignments for the DS1620.

By configuring the DS1620 for CPU mode, you will enable the 3-wire communications interface. This interface is serial—unlike RS-232, the 3-wire communication is synchronous. To start communication, the select line, RST, goes high to enable the chip. The clock line, CLK, is brought low, and the first bit (least significant bit) of the byte is placed on the data line, DQ. CLK is then set high; it is upon the rising edge that the data is read into the DS1620. CLK will then go low. The next bit is then placed on DQ, and the process is repeated for all eight bits. Reading from the DS1620 is similar except that the microcontroller will sample the DQ line after each low-to-high transition of CLK.

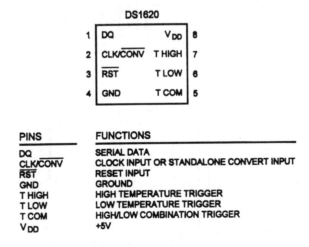

Figure 11-4. A pinout for the Dallas Semiconductor DS1620. This temperature sensor can operate in stand-alone mode or when interfaced with a microcontroller.

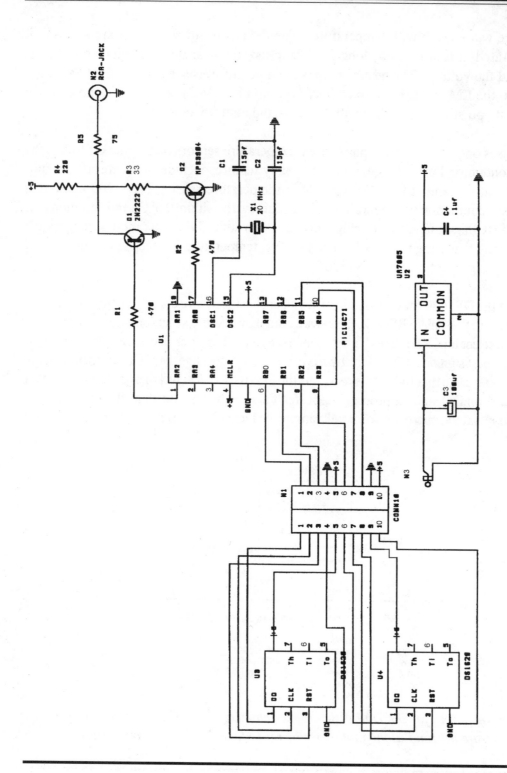

Figure 11-5. *A schematic for the video thermometer:*
Hard to believe, but this circuit will produce an NTSC-compatible video signal.

To read the temperature, you must first send the 8-bit command AAh. After the last bit of the command has been clocked, reverse the I/O direction of the microcontroller pin. The microcontroller will proceed to issue clocks on the CLK line, reading each bit of the returning temperature number. The problem of sensor cable length has been mentioned many times, and the digital reading of the temperature is precisely why the DS1620 is not subjected to it. Just like in the old argument of CD player vs. record player (remember vinyl?), digital information is not subject to the same kind of loss as an analog signal.

Circuit Description

The schematic for the video thermometer is shown in *Figure 11-5*. Yes, this is the entire schematic. The power of the PIC® 16C71 now becomes more evident. The PIC® 16C71 has 13 I/O lines of which four can be inputs to the internal A/D (not used in this project). It has 1024 bytes (14-bit wide) of program memory and 36 bytes of RAM. The PIC's® interrupt on the internal timer presents an easy method for generating the horizontal timing pulses. Incorporated auxiliary features (power-on reset, oscillator circuit, watchdog timer) allow for the single-chip aspect of the project. Crystal X1 is 20 MHz, producing 5 MIPS (million instructions per second) from the PIC®.

The I/O line RA2 drives Q1, dropping the video line down to 0 volts for horizontal and vertical sync. RA0 drives Q2, pulling the video line down to 0.25 volts for the blanking level. If neither Q1 nor Q2 is on, the video line will be one volt (white video pixel). Resister networks R3, R4, and R5 generate the proper voltage levels into a 75-ohm load. Notice that

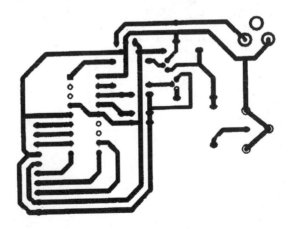

Figure 11-6. The board layout for the video thermometer.

I/O line RA1 is tied to ground. This is important because of the way the video pixel information is created. To achieve maximum speed (and therefore the smallest pixel), you will rotate each bit into RA0. RA0 reads the I/O port before the bit shift, and if RA1 is floating high, it will shift a high to RA2, turning on the sync transistor. By grounding RA1, you can guarantee that there will be no spurious sync signals. See the *firmware* listing for more on this method of video generation.

The temperature sensors are directly connected to the microcontroller through N1. Power is provided to the DS1620s via N1. A regulated five volts is provide by Q3, a 7805. Board power could be from a battery, but for long operation you should use an AC-to-DC wall adapter; all in all, an amazingly compact device for what it does.

Construction

This board could easily be wire-wrapped, or you can cut your own PCB using *Figure 11-7* for parts placement and *Figure 11-6* for circuit board trace layout. Because of the simplicity of this project, little can go wrong. The cable to the sensors can be up to one-hundred feet long—be sure to use a thicker gauge wire if you do have a long cable run. The board is designed to run continuously, so no ON-OFF switch is provided; however, be sure to add one if you plan to use a battery to power the board. *Figure 11-8* shows the completed video thermometer.

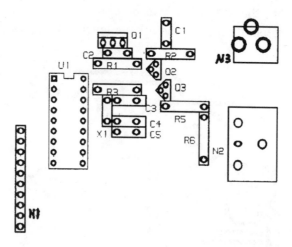

Figure 11-7. Part placements for the PCB layout in Figure 11-5.

Figure 11-8. The completed video thermometer. Small, compact, yet very powerful.

It was mentioned earlier that the video generated may be too intense for some TVs or monitors. If this is the case for you, adjust resistor R4. Increasing the value will decrease the brightness of the signal. A potentiometer could be added here as a brightness control. Be sure to leave R4 in series with the pot, or too much current could flow through Q1 and Q2.

To enclose the project, a simple, small plastic case will do. You need to drill only three holes for the temperature sensors, power-in and video-out cables. If the temperature sensor will be outside of the case, seal it against the weather with heatshrink tubing.

Operation

Plug in the power supply, place the sensors where desired, connect the thermometer to the video input jack of your TV or monitor, and it's working! The firmware listed here will produce the words INSIDE and OUTSIDE, designed for one indoor sensor and one outdoor sensor. Changing the wording of your setup is easy—simply modify the firmware. However, keep in mind that you must use the same number of letters per line, or a major rewrite will be necessary.

The DS1620 reports the temperature in Centigrade. In the United States, people are more familiar with the temperature in Fahrenheit. The PIC® automatically translates the temperature reading to Fahrenheit. Should you desire Centigrade, rewrite the firmware to skip the translation process; or, how about a modification to alternate between displaying Centigrade and Fahrenheit?

Most modern TVs have multiple input jacks for a number of sources. You can plug the video thermometer into an open video input and instantly switch to the temperature reading with a flick of the remote control. With PIP (picture in picture), the temperature can be seen simultaneously with regular programming. What if your TV does now have any video inputs at all? Purchase a readily-available channel 3 RF modulator, and pipe the video into it. Now the reading will always appear on channel 3 of your TV.

Firmware

Most of the work in this project is done in firmware. To help clarify the code, a flowchart of the program is presented in *Figure 11-9*. After we discuss the flowchart, we will see a few highlights of the firmware, with the emphasis on routines unique to the PIC® 16C71. The entire code is listed in Appendix I.

From the flowchart, you can see the program starts at *power up*. After initializing both the temperature sensors and the internal PIC® registers, you then proceed into the main program loop. The loop performs three tasks per sensor; read the temperature, convert it from Centigrade to Fahrenheit, and convert the reading into three digits stored in the PIC® registers. These jobs are constantly running outside of the time required for the interrupt-driven video generation. Follow the main program loop, and you will see how straightforward it is, especially when compared to the befuddlement of this interrupt routine.

After the initiation of the timer, you will start to receive timer interrupts every 63.5 µs. The video signal can be roughly divided into three portions: vertical sync, equalization lines and video lines. At the beginning of the interrupt, test to see which segment is being processed. Each segment has its own "run" flag and line counter, which you can test to see if the segment is done (3 vertical, 19 equalization, 240 video — a total of 262 lines for one video frame). Both the vertical and equalization lines create pulses and then return from interrupt. The 240 lines with possible pixels start with the 5 µs horizontal pulse. Then, test for the end of the segment and the flag for processing characters on the screen. When the flag is set, the program loads the screen buffer with the pixel pattern for the current row of characters in the

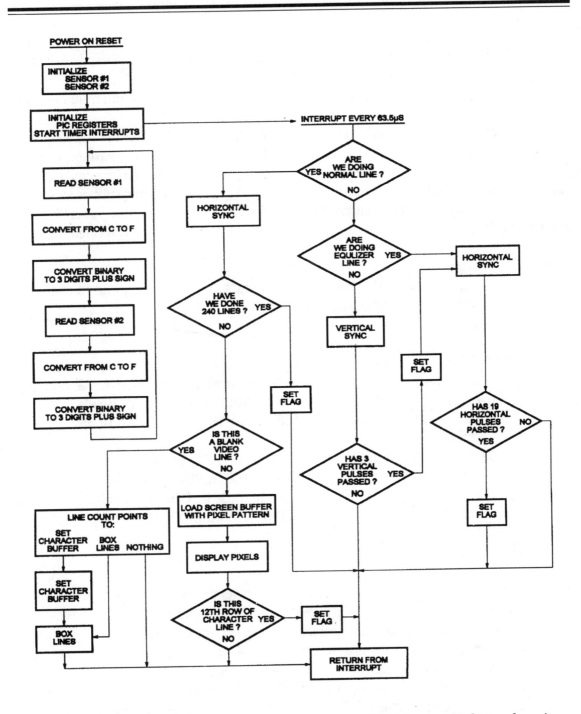

Figure 11-9. *The firmware is the core of this project. To understand it better, here is a flowchart. Please reference the flowchart as you read through the firmware.*

character buffer. The display appears after the screen buffer is loaded. After the display, test for the end-of-character line, and return from interrupt. If the processing character flag is not set, test the line count to see if you need to generate the "box," load the character buffer with new information, or do nothing. After performing their tasks, they all return from interrupt.

Now, let's look at some of the highlights of the firmware. Unlike PICs® of the 16C5X family, the PIC® 16C71 will start execution at location 0. The interrupt vector is at location 4. There obviously cannot be much of a program between locations 0 and 4, so the first command is a jump to the location of where the real program starts. The interrupt could start processing at location 4, but there is a need to utilize the first 256 bytes for a subroutine. Therefore, the interrupt vector is also a jump to where the real interrupt handling code is located:

```
            org     0
            jmp     start   ;PIC16C71 always start at location 0, jump to real start
            org     4       ;interrupt address
            jmp     handler
        ;
    get_matrix              ;get pixel pattern for character
            add     w,row   ;enter with character number in w (ex: 1 is 12, 2 is 24)
            jmp     pc+w            ;then add current screen row for pixels
            retw    0c3h,81h,00,18h,18h,3ch,3ch,18h,18h,00,81h,0c3h   ;0
            retw
0c3h,0c3h,0e3h,0e3h,0e3h,0e3h,0e3h,0e3h,0e3h,0e3h,00,00   ;1
            retw    0c3h,81h,18h,18h,0f1h,0c3h,87h,8fh,1fh,1fh,00,00   ;2
            retw    0c3h,81h,18h,0f8h,0f8h,0c1h,0c1h,0f8h,0f8h,18h,81h,0c3h;3
            retw    18h,18h,18h,18h,00,00,00,0f8h,0f8h,0f8h,0f8h,0f8h   ;4
            ............
            ...........
```

The subroutine *get_matrix* is a table lookup for the pixel pattern of the current row of the current character. In the register *row*, you have the current row number (1 to 12) of the character. Before calling this subroutine, move the character number into the work register (w). The character number is in multiples of 12. '0' is 0, '1' is 12, '2' is 24, '3' is 36, and so forth. By adding the row register to the character number you have the number of bytes to jump into the table to retrieve the pixel pattern. The next instruction, *jmp pc+w*, simply adds the work register to the current value of the program counter (the first byte of the table). Thus, if the work register is 40 and the program counter is 7, the next instruction would be executed at location 47. Each byte of the table executed at location 47. Each byte of the table is a return from the call instruction, which also places a designated value into the work register. Therefore, you return from this subroutine with the pixel pattern in the work register.

One last note on the *jmp pc+w* command: the program counter is 14 bits long, with the high-order bits in a separate register from the low-order bits. Adding the work register is an 8-bit operation with no automatic carry and increments of the high-order bits. What does this mean? If the program counter contained 200, and the work register 100, then the result of the addition would be 44 (300 - 256 rollover). The next instruction would be executed at location 44 instead of the desired location 300. This is why the subroutine *get_matrix* is placed near the start of the program page:

```
;
handler                         ;interrupt routine
        mov   w_copy,w          ;save copy of work register
        mov   s_copy,status     ;save copy of status register
        clrb  T0IF              ;clear timer interrupt flag
        mov   TMR0,#sync        ;ready for next interrupt
.............
.............
;
h_exit
        mov   status,s_copy     ;restore status register upon exit
        swap  w_copy            ;from interrupt
        mov   w,<>w_copy  ;swap work register into w, status is unaffected
        reti
```

As you know, an interrupt can happen anywhere in the body of the main program. The program could be in the middle of performing a task, depending on the values of the work and the status register. The interrupt routine will change the content of these registers. Upon returning from the interrupt, the previous executing code will have the wrong values. To correct this, move both the work and status to temporary registers at the start of the interrupt (see *handler*). The final instructions of the interrupt restores the work and status from those temporary registers:

```
        mov   fsr,#screen_buf;prepare for indirect addressing of screen buffer
        clrb  blank               ;line before character for box
```

```
            mov    col_count,#7              ;number of bytes in screen buffer
            nop                              ;need for width of line
            setb   blank                     ;end line before character
next_char                                    ;we will now create pixels on screen
            rl     indirect                  ;indirect is pointing to pixel screen buffer
            rl     ra                        ;rotate left buffer, pixel goes to carry bit
            rl     indirect;rotate left I/O port, carry sets high or low on video line
            rl     ra                        ;this way a pixel is set in two instructions
            rl     indirect                  ;or 0.4 us
            rl     ra                        ;this is also why we grounded pin RA1
            rl     indirect                  ;to prevent rotating a high into the sync
            rl     ra                        ;pin RA2
            rl     indirect                  ;no loop on this code to save time
            rl     ra
            rl     indirect
            rl     ra
            rl     indirect
            rl     ra
            rl     indirect
            rl     ra
            inc    fsr                        ;next screen buffer
            setb   blank                      ;black after character
            djnz   col_count,next_char        ;time for space between characters
```

The code at *next_char* uses indirect addressing and rotation to quickly generate pixels onscreen. Before the start of the current line of pixels, *screen_buf* was loaded with the correct pattern. The register *FSR* (the index register for indirect addressing) is loaded with the address of *screen_buf*. In indirect addressing, the command *rl indirect* will work on the register that *FSR* is pointing to. Thus, if *FSR* contains 10, then *rl indirect* will rotate to left register number 10. The pixel is rotated into carry, then carry is rotated to the video line. This method can produce a pixel 0.4 µs in width. After eight pixels, you increment FSR to the next character. This is repeated seven times for each character:

```
                        ;prepare PIC registers
    setb   rp0          ;switch to register page 1
    clr    wdt          ;assign prescaler to RTCC
    mov    option,#0    ;prescaler divide by 2
```

```
mov    adcon1,#3    ;configure port A as non-A/D
clrb   rp0          ;switch to register page 0
mov    TMR0,#sync ;start sync time delay
mov    intcon,#10100000b ;interrupt register set for only RTCC interrupts
```

At the start of program execution, you initialize various PIC® registers. In the PIC® 16C71, some registers can only be accessed by first setting a register page flag. This is the *setb rp0* command. The interrupt is driven by the time out of TIMER 0. You assign a prescaler (divide by 2) to TIMER 0 to generate the correcting timing pulse of 63.5 μs. The value *#sync* is moved into TIMER 0 at the start, and on each interrupt, to continuously generate the horizontal sync pulse.

CHAPTER 12
Practical Design Considerations

No doubt, dozens of microcontroller-based applications are now sprouting in your fertile imagination. The fun part of electronic design is the conception and construction of the initial prototype. As a design *engineer*, you must then do the often unpleasant but absolutely necessary task of transforming your creation into a product that is practical to manufacture and reliable for the end user. If you point-to-point wire a prototype for motor control, how will the equivalent PCB traces handle the heavy current load? In producing a one-off, you may use expensive components. Is it economically feasible go into production with such an item? Is the production design compatible with all federal laws? These are not fun problems to work with, but before you put the better mousetrap (*with a built-in microcontroller*!) on the market, it is essential to cover all the bases.

What are some of the factors you will be looking at? Of lesser importance but still helpful is the way you built the prototype. Creating a PCB prototype verses a wire wrap will provide valuable insights to the way the production unit will function. Of course this is directly a result of the PCB layout: size, speed, power and form must all be considered. As mentioned in the opening pages of this book, you can't add a $20.00 electronic gadget to a $15.00 coffee pot. Oftentimes, a working prototype is redesigned to lower cost; this is frequently referred to as *value-engineered*. In the "real world," your product will encounter a number of predators. You should design the item to be rugged and immune to *electrostatic discharge* (ESD). Is our item susceptible to radiated emission from other electronic products? Does your unit produce emissions? A production item must meet FCC limits on *electromagnetic compatibility* (EMC). For the international market, the European Community (EC) has established new and strict rules for both EMC and ESD. Are you in compliance?

Prototype

There are three ways to prototype a board: point-to-point, wire wrap, and PCB. Point-to-point uses a board predrilled on a .1 inch matrix, with or without pads. These boards are called plugboards. For microcontroller applications, plugboards with power buses are recommended (long connected strips of holes), and connected pads based on a standard DIP socket width. In all three basic prototype methods, because of unavoidable mistakes, use sockets for most ICs. This will save you much time later. Solder the sockets and other parts to the plugboard. Connect the parts by following the schematic. Use 22- or 24-gauge wire unless heavy current draw is expected, then used 16-gauge wire or whatever is appropriate. Advantages: quick, low cost, and capable of applications with heavy current draw. Disadvantages: error prone, not a good model for production, and unsuitable for large prototypes.

Wire wrap is used primarily for digital prototypes. Special sockets are soldered to plugboards with power buses. These sockets have long, square pins as leads. Then, 30-AWG wire is "wrapped" around the sockets pins following the route of the schematic. There are special tools to wrap the wire. These tools cut, strip and wrap the wire in one quick operation. Some are even motorized. Since wire wrap sockets are DIPS, how do you handle analog components? You can purchase "headers" that plug into the sockets. To these headers, you can solder capacitors, resistors, transistors, etc. Advantages: quick large digital prototypes. Disadvantages: high cost, line cross-talk, analog "unfriendly," and only good for low current.

Directly cutting a PCB is your third and final option. Let's immediately look at the advantages and disadvantages. A PCB prototype is excellent for both digital and analog, high or low current. As a model for production, it is (of course) unsurpassed. The size of the prototype is limited by in-house facilities. Most private companies and students should easily be able to do 8 x 8-inch boards. What are the disadvantages? The initial cost of the equipment is high. Only single- and double-sided boards are practical. Persons requiring large or multilayer prototypes should go to a board house for a prototype.

With the advances in producing your own PCB, this is the recommended method when compared to the other options. Not too long ago, cutting a PCB involve many time-consuming steps. First you had to lay out a pattern on paper with special scaled tape and pad patterns. Then you photographed this pattern and produced a negative to scale. Using light, you transferred this pattern to a pre-sensitized copper board. Then you etched and drilled.

With your home computer system, (most electronic students and hobbyists have access to a PC), cutting a PCB is quick and easy. First, the pattern is created one of two ways. An electronic CAD package can be used. One low-cost but very powerful tool is the WINboard from IVEC. For around $30.00, this software will design PCB layouts up to 200 pins. The second method can only be used with a predrawn layout, such as the boards presented in this book or layouts in the magazines *Popular Electronics* and *Electronics Now*. A low cost (under $50.00) black-and-white hand scanner can produce excellent copies of any printed PCB layout. A number of drawing software packages can modify the layout or rescale the drawing if needed. The pattern is transferred to a special plastic film with a laser printer. Two different brands of film are TEC-2000 and Toner Transfer System. Then, using an iron, press the pattern onto a copper board. Etch, drill, and you are done!

What if you don't have a computer or laser printer? If the pattern is to scale, take the pattern to a photocopier and copy the layout onto the special plastic film. Some copiers can reduce or enlarge, but scaling is difficult to do with these units. All of the projects in this book can be created with the direct PCB prototype method.

PCB layout

Again, we approach a subject about which entire books have been written. We will skim the surface of this topic to illustrate a few of the items you should be aware of. The layout of the traces on a PCB is a major factor in fighting ESD and EMC. We will cover these aspects a little later. In an average microcontroller application, the microcontroller is the heart of the board. Inputs flow to the controller and outputs radiate from it. Start the PCB layout with the microcontroller in the middle. Place connectors or components close to their respective I/O pins of the microcontroller. Keep high speed, analog, and high current lines to minimum length. Professional board houses can obtain 5 *mil* lines and 5 *mil* spaces, but this is nearly impossible with home-cut boards. Use at least 20 *mil* lines and 10 *mil* spaces. Pads should be as large as possible. Don't be afraid to use elliptical pads; this gives the largest area while maintaining space between pins for traces.

Decoupling caps are important for fighting EMC and for the correct operation of certain chips. Always place these caps (*typical .1uf*) as close to the power pins of the chip as possible. One cap per high-speed digital chip is the rule of thumb. There have been cases of entire boards failing to operate because of missing decoupling caps. Of course, there should also be a primary filter cap on the power supply.

Separate line AC from the rest of the board. Look at the layout of the motor control project in Chapter 9, *Figure 9-4*. You can clearly see a blank zone starting at the transformer, and wrapping itself over to the optocouplers and down the middle of the optocouplers. This separation is an important safety factor.

A primary element to consider is the current draw of a PCB trace and the required width to support it. Copper boards come in different thickness (1 oz, 2 oz, etc.), and this plays an important factor in heavy current applications. *Figure 12-1* is a chart showing the temperature increase on a trace verses the copper weight and width of the trace. To use this chart,

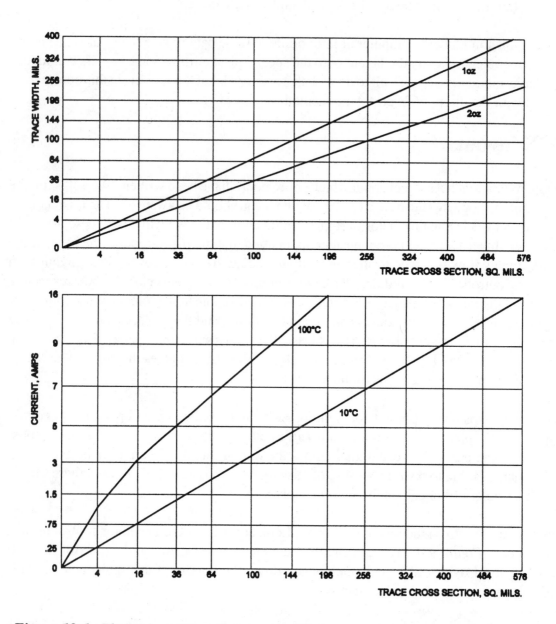

Figure 12-1. *These charts show the temperature increase (over ambient) for the current flow in different-sized (width and weight) copper traces. First, use the top chart to find the area of the trace cross section. Use that number with the expected current draw on second chart to find the temperature increase.*

first find the area of a cross section of the trace. A trace .025-inch (25 mil) at 1 oz copper would be 36 square mils. Now, move up the chart and see the temperature rise as the current on the trace increases. 1.5 amps would rise the trace temperature by 10° C over ambient, and 5 amps would increase trace temperature by 100° C! With too much current on a trace it will pop like a fuse. Slightly excessive current will cause the trace to turn black and eventually peel off from the board.

Cost

In a production environment, rarely do you have the luxury of allowing any waste on a final product. The microcontroller should have just enough power to do the job required and no more. Extra I/O lines, extra memory, and unused capabilities (A/D, interrupts) all cost money. If the prototype is functioning perfectly and the controller has excess capacity, see if a lower chip in the family can do the job. It is not uncommon to produce thousands of a microcontroller application in a production run. If you can reduce the cost of the controller by just fifty cents over fifty thousand units, the savings would be $25,000. Well worth the effort. In the next chapter, there is a chart showing the differences between microcontrollers and controllers within the same family. See if one can the do the job at a lower cost.

The next recommendation is strictly for an industrial environment . You could select a different microcontroller for each application. However, consider the cost of maintaining the inventory of a number of different chips, the increased cost of low volume purchasing (verses large volume single chip purchasing), and the development tool cost for a number of different controllers. These cost factors are often unseen but very real. Weigh the cost increase of different controllers verses the savings that a precise matching of capabilities will bring, and choose the most effective course.

All components on a board should be rated at the appropriate voltage, current, and speed. Any excess will cost more than required. The power supply should be carefully matched to the constant current flow, and have the ability to provide temporary peak current flow. If you can meet this criteria, substantial savings can be realized. Now for the *don'ts*. First of all, never underrate capacitors to lower their cost. Plus, there should be absolutely no skimping on circuits dealing with high current or line AC. There are no cost savings when your products fail to operate.

ESD

ESD is a killer of chips and products. You must design your products to withstand or prevent ESD. First, consider ESD in the manufacturing environment . Simply passing a microcontroller from one person to another will create a discharge. Bear in mind that an ESD event below a few thousand volts cannot be felt by people, but at that level, there can be damage to a semiconductor device. Worst case scenario is that a device will not be immediately bad but will deteriorate with time and produce a failure in the field. Entire product lines can become unreliable because of poor manufacturing procedures. A few quick manufacturing rules: always keep chips in antistatic carriers. Production personal should always wear ground straps, and soldering stations should be well grounded. Boards under construction must be transported in antistatic trays. There should be absolutely no passing of unprotected chips or boards between people.

When the product is in the field, there are three primary types of ESD events. First is a direct ESD hit on the input of a device. Example: a discharge from your finger to a keyboard, or from the keyboard to the input of a microcontroller. Next is an ESD ground bounce. Let's say your PCB has a ground which is connected to the case of the product (a common occurrence). Some grounds have low (but not zero) impedance. The quick rise time of an ESD event (*typical 1 nsec*) will cause the ground to "bounce." This will not destroy devices, but will cause CMOS circuits to latch-up, potentially causing severe damage in the operation of the device. Thirdly, indirect coupling of an ESD event. If the case has an excellent ground, it will prevent a direct discharge to the PCB. However, the very action of discharging into a ground will produce an electromagnetic field, which can create a smaller shock wave in the PCB by inductance.

What can be done? Look at the product as a whole. Where is contact (and therefore possible ESD events) most likely? Ground these areas thoroughly, or insure that they are nonconductive and not even close to internal PCBs, cables or wires. A good example of this is a metal door on the printer of a computerized cash register. Example: this door is completely surrounded by plastic and isolated from ground. Every time a clerk discharges to the metal, the CPU board receives an indirect hit and locks up. Grounding the cover solves the problem. Any leads from the PCB to the outside world should be as short as possible, and protected with filters (or buffered with a low value input resistor). On larger PCBs, consider multilayer boards with ground planes. These can be up to ten times as immune to an ESD hit. Finally, add software protection. This is a good use of the watchdog timer in a microcontroller.

When a controller receives an ESD hit, it will typically lock up. The watchdog timer times out and resets the controller. In critical applications, always enable the watchdog timer.

FCC & E.C.

The FCC is interested in preventing RFI (radio frequency interference). To this end, they have established emission regulations called *FCC part 15*. Any device which has digital circuity operating in excess of 9 KHz must pass *FCC part 15*. The regulations are way too lengthy to list here; just keep in mind that the FCC will test for excessive conductive and radiated emissions in a number of frequency bands up to 2 GHz.

What can you do to limit emissions? Keep the PCB trace lines as short as possible, especially clock lines. The liberal use of decoupling caps and ground planes will also help. A well-grounded case and shielded cables are invaluable to limiting RFI. Does this sound like some of the recommendations to prevent ESD? Indeed, good RFI prevention will limit ESD, and good ESD prevention will limit RFI.

The EC has instituted new and strict rules for both EMC and ESD. That's right: not only must your products restrict emissions, but they must pass immunity tests against ESD. Following the above rules on PCB and case design is a good start toward meeting regulations. If you are designing a product for the USA or EC, please read one of the several books available on EMC regulations.

The EC has one other requirement which is currently not necessary in the USA: the EC will test for product immunity against power line disturbances. The product must withstand "brownouts" to a certain voltage level. Also, the device must have immunity to power line transients. Consider all these factors before designing a product for the consumer market.

CHAPTER 13
Other Microcontrollers

At the beginning of this book, we talked about what constitutes a microcontroller. We also discussed several families of controllers, touching briefly on similarities and differences. We then started with the fundamentals behind controllers, digital electronics and Boolean logic. At this junction, it was necessary to focus on one family of microcontroller since it would have been unrealistic to detail the internal workings and instruction sets of dozens of controllers. After examining the Microchip PIC® line, this knowledge was put to use in both hardware and firmware in five projects. You should feel confident that you are now ready to design your own microcontroller application with a PIC® as the controller.

But what about other microcontrollers? The knowledge you have accumulated to date is easily transferable to all digital microcontrollers. The only devices which are quite different are analog and "fuzzy" microcontrollers. In fact, it will serve you well to consider the multitude of controllers available before settling on one particular controller for your application. If you are a novice to the field, how do you know what's available and the capabilities of a device? We shall review many microcontrollers shortly.

After a general review of different microcontrollers, we will consider a number of controllers in greater depth. Some of these controllers are above 8 bits. While this book has concentrated on 8-bit microcontrollers, you should be aware of these powerful devices which you might need for a difficult task. We will also look at expanding a typical microcontroller, the 8051. Lastly, we will revisit writing firmware with a longer look at using a C compiler. Compare these C code examples to the Assembly language of our previous projects.

Microcontroller Differences

Cataloging all of the available microcontrollers can be a daunting task. In addition to the number of different manufacturers, each microcontroller potentially has scores of variations. Lets consider the 8051. At last count, there are 98 versions of the 8051 produced by 8 manufacturers. As previously discussed, what makes these versions different are the internal modules such as timers, RAM, EEPROM, A/D, etc. The 8051 is unique because some of the variants are 16-bit or have modifications to run at exceptionally high speeds.

Table 13-1 is an attempt to list and compare many of the microcontrollers on the market today. Only a few representative controllers from one family are listed, giving an idea of the capabilities of the chip. The microcontroller name can be specific (Z89120) or the family name (Z8). The RAM and ROM/EPROM columns give the largest available value for the chip or family. The clock value is also the highest rating for the line; usually, there are slower versions for less money. MIPS is very hard to calculate because most microcontroller instructions vary in execution time depending on the type of instruction. Therefore, a rough average was used to rate MIPS. I/O pins, PWM, A/D, and interrupt often share pins. The maximum was listed for each option; however, bear in mind that you most likely cannot have the maximum of each feature simultaneously. The price is extremely variable, depending on many factors such as quantity, distributor, and current demand. With reservations, it is as a means of giving you a rough ideal of the comparative cost. Any figure with a star (*) by it has numerous possibilities, and there is no way of coming up with an average number. Finally, any information could not be found at present was left as a blank field. With these caveats been said, please look at *Table 13-1* as a guide to many of today's microcontrollers.

8051

The "Model T" of microcontrollers, the 8051, has been with us since 1980. Intel was the original manufacturer; several others, including AMD, Philips/Signetics, Matra, and Siemens, have been licensed to produce the chip. The basic 8051 comes in a 40-pin DIP package, has 128 bytes RAM, 4K bytes ROM, and 32 I/O lines. While the 8051 has an internal mask ROM (with a fixed program), you can disable it and run from external ROM or EPROM. This is like the 8031, which has no internal ROM. The 8751 is the same as the 8051 except that EPROM replaces the ROM, making it perfect for developmental purposes. The 8051 can also access external RAM, greatly increasing the data storage area. Both external ROM and RAM can be as large as 64K. To implement this, I/O lines are used for the address and data bus; however, using external RAM is not as efficient as internal RAM.

Variations of this microcontroller are truly impressive. The first 8051 had a 12 MHz clock, 12 clocks to one internal cycle, and multiple cycles per instruction. This produced a rough MIPS of .5. High-speed versions (DS80C320, DS87C530) are on the market today, which offer clock rates up to 33 MHz plus improved clocks to cycle time, generating up to 10 MIPS. The ATMEL 89C1051 is a 20-pin chip with FLASH memory, 2 timers, UART (serial), and low power. The DS5000 is a standard 8051 with nonvolatile RAM in the place of ROM—perfect for changing application programs on the fly. Philips produces an 8051 variant that seems to compete with PICs® for the low-end market: the 83C750 (approximately

Microcontroller	RAM	ROM/EPROM	Clock	MIPS	Instructions	EPROM	I/O Pins	Timer	Serial Ports	Watchdog Timer	PWM	A/D	Interrupt	Price
37700	2K	32K	25Mhz	6	103	No	68	8	Yes	Yes	Y	Y	3	9.50
68HC05	1.2K	16K	8Mhz	1	62	No	32*	2	Yes	Yes	Y	Y	Yes	2.75
68HC11	1.2K	32K	16Mhz	2	62	512	38	2	Yes	Yes	N*	Y	3	7.94
68HC16	2K	48K	4Mhz	8		Yes	99	4	Yes	Yes	Y	Y	7	14.34
78356	2K	32K	32Mhz	5		No	54	5	Yes		N	Y	Yes	14.00
8051	128	4K	12Mhz	1	46	No	32	2	Yes	No	N	N	2	3.50
80C166	2K	32K	40Mhz	10	76	Yes	81	5	Yes	Yes	4	Y	38	26.00
AT89C2051	128	2K	24Mhz	2		Yes	15	2	Yes	No	N	Y	2	2.00
COP820CJ	64	1K	10Mhz	1	55	No	23	3	Yes	Yes	2	N	3	2.50
COP888EG	256	8K	10Mhz	1	55	No	39	3	Yes	Yes	3	N	2	5.80
DS80C320	1.2K	4K	33Mhz	10	46	No	32	2	Yes	Yes	N	N	6	
H8	2K	62K	25Mhz	5	63	Yes	58	8	Yes	Yes	2	Y	9	11.25
MCS 96	1K	32K	20MHz	5		No	56	2	Yes	Yes	Y	Y	Yes	13.85
PIC16C54	32	512	20MHz	5	33	No	12	1	No	Yes	N	N	No	2.34
PIC16C55	32	1K	20MHz	5	33	No	20	1	No	Yes	N	N	No	3.20
PIC16C57	80	2K	20MHz	5	33	No	20	1	No	Yes	N	N	No	3.75
PIC16C64	128	2K	20MHz	5	35	No	33	3	Yes	Yes	1	8	8	5.00
PIC16C74	192	4K	20MHz	5	35	No	33	3	Yes	Yes	2	8	12	6.25
PIC16C84	36	No	10MHz	2.5	35	1K	13	1	No	Yes	No	No	4	4.65
PIC12C508	25	512	4Mhz	1	33	No	5	1	No	Yes	No	No	No	1.88
PIC12C509	41	1K	4Mhz	1	33	No	5	1	No	Yes	No	No	No	2.05
ST6211	64	2K	8Mhz	0.3	28	No	12	1	No	Yes	No	No	Yes	1.50
ST6293	128	4K	8Mhz	0.3	28	128	20	3	Yes	Yes	No	Yes	Yes	2.70
ST9026	256	16K	24Mhz	2	87	No	40	1	Yes	Yes	No	No	Yes	6.90
TLCS90	1K	32K	16Mhz	4	163	Yes	54	5	Yes	Yes	Yes	Yes	4	9.55
TMPN3120	1K	10K	10Mhz	2	74	512	11	2	Yes	3	1	No	No	10.00
TMS370	1K	32K	20MHz	5	73	512	55	2	Yes	Yes	Yes	Yes	3	8.85
Z8	256	20K	20MHz	10	46	No	56	2	Yes	Yes	No	No	4	6.70
Z86E33	237	4KB	12MHz	6		Yes	24	2	No	Yes	No	No	Yes	
SX18AC	136	2K	50MHz	50	43	Yes	12	Vir	Virtual	Yes	Vir	Vir	Yes	3.00
SX28AC	136	2K	50MHz	50	43	Yes	20	Vir	Virtual	Yes	Vir	Vir	Yes	3.50

Table 13-1. A sample of the microcontrollers available on the market today.

$1.00 in high quantity). Siemens 80C517a is highly integrated with a high speed 8051, large memory, multiple UARTS and PWMs, and a 32-bit accumulator. Both Intel and Philips have produced a 16-bit version of the 8051. Intel's chip is pin- and binary code-compatible with the old 8051, making upgrades easy. The Philips version is source code-compatible. This means that you have to recompile the original source code and create a new binary file in order to program the chip.

Address Decoding

We have covered basic microcontroller design with many of the projects in this book. Let's now look at an advanced design using the 8051 as the core controller. Because of the size of the application, it is necessary to use external RAM and ROM. We will also attach several peripheral chips, such as UARTs, parallel printer ports, video display, and general-purpose I/O lines.

Warning!
This next section will be difficult for the novice microcontroller user. It is not possible in this book to cover all of the nuances of a common bus design. The purpose is to show you the possibilities surrounding an extreme controller design. After you have done some of your own designs, reread this section to enhance your growing knowledge of microcontrollers.

While microprocessor design is not the focus of this book, you must understand some aspects of it since the peripheral chips were intended to be used with microprocessors. A microprocessor communicates with external chips by address and data lines. Address lines contains a binary number which locates the memory or I/O byte that the processor is currently working with. The data lines contain the byte value coming from or going to the processor. Depending upon the processor, you then have various clocks and strobes which synchronize the data transfer. An example would be the write (WR) strobe, which toggles to write a byte to memory or I/O port.

Memory and I/O chips have one or two chip select lines, multiple address lines, and eight data lines. As an example, let's say that you have a memory chip that stores 256 bytes. To address each location, this chip needs eight address lines. To read and write data, you need eight data lines. You also need chip select, write, and read lines. Since you have address lines, what does the chip select line do? Let's extend the required memory to four of these chips. All of the data, address, and clock lines are tied together in parallel. By setting the chip select line, you can address each chip separately, which is necessary in order to read or write bytes to the individual chips. This problem is multiplied when you realize that all of the memory and I/O chips are on common address and data lines.

This is where address decoding comes in. Each chip needs a separate chip select line. By using TTL chips, you can break the address into defined areas for either memory or I/O. *Figure 13-1* is a partial schematic of a large 8031 application. In it, you can see the address decoders and separate chip select lines. U4 is a one-of-eight decoder. The three most significant bits of address lines are fed into U4. The output is one of eight lines, which is selected based upon which 8K (64K / 8 - 8K) address block the processor is currently addressing. This is further defined for I/O chips by U3, a one-of-sixteen decoder. If you are in the forth 8K address block, then this chip is active and selects its output based upon the 128 byte block that the processor is currently addressing. (Notice the 74LS154 has A10 as the most significant bit; 10 address bits = 2048 bytes; 2048 / 16 = 128 bytes.) This dividing of the memory space is shown in *Figure 13-2*. As you look at *Figure 13-2*, keep in mind that the 8031 has two 64K blocks of memory: one for code (ROM) space, and one for data (RAM) space. All of the divisions are in the data space.

The 74LS154 has 16 output lines; each one can be connected to an I/O device. You could connect five serial ports, five parallel printer ports, and six general I/O chips (144 I/O lines) to your microcontroller! To reference them in the firmware, simply treat the external chip's registers as RAM locations in the data space. Example: if an UART was in I/O location 1, then the data register of the chip could be reached by using address 24576 in any acceptable command in the 8031.

The 8031 uses 16 external address lines, which can address a maximum of 64K external memory; however, just to make life more difficult, the 8031 does *not* use 16 of *its* pins for addressing. In order to save I/O pins, the low order address byte (A0 - A7) is multiplexed with the data lines. To separate the address from the data bits, an 8-bit latch is used. In *Figure 13-1*, this latch is U2, a 74LS373. During an external fetch cycle, the address lines are set first and latched by use of the ALE line. After the address is out, then the lines revert to data for input or output.

Figure 13-1 also shows an external EPROM (U22) connected to the 8031. Some peripheral chips require an I/O read/write line. An OR gate, U5, generates this signal from data read

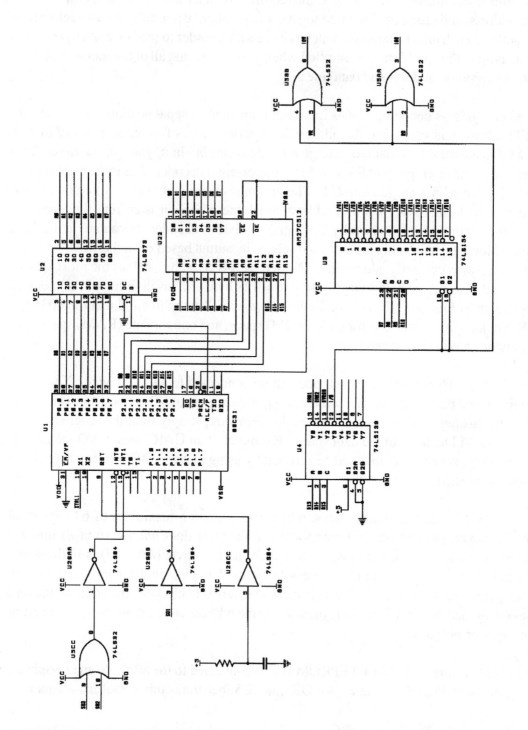

Figure 13-1. An 80C31 microcontroller with an external EPROM and address decoding. The clock for the 80C31 (XTAL1) comes from sheet 2 and the simulated wait state circuitry.

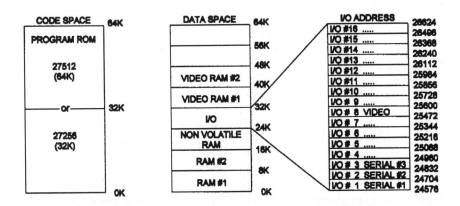

*Figure 13-2. This memory map shows how memory is partitioned
by the address decoding in Figure 13-1.*

RD, data write WR, and the I/O select line. There are three possible sources of external
interrupt, and only two external interrupt input pins on the 8031. To handle this, place the
interrupt with the highest priority on INT0. The two remaining interrupts are ORed (U5)
together and sent to INT1.

Adding a Video Display

In your last project, you saw a microcontroller produce a video display. While useful in that
application, what if you need to free the microcontroller's time to deal with dozens of I/O and
communication lines? Also, you were very limited in the number of possible characters. Can
you add a video display with 80 character lines by 24 rows? *Figure 13-3* is a continuation of
the schematic in *Figure 13-1*. The heart of the video section is the NCR 72C81, CGA generator. The chip was designed to work with an IBM PC bus; its addressing and I/O ports are
compatible with a standard IBM CGA card. The entire video signal is generated by this chip,
greatly reducing the load on the 8031. In fact, to display a word, the 8031 has only to write
the character to memory (*Figure 13-2*, Video RAM), and it appears on the screen.

The 72C81 has 20 address lines while the 8031 has only 16. U31 and U32 (74LS244) changes
the 8031 16-bit address lines to the expected 20 lines for the 72C81. 8031 memory location
8000-B000 becomes B8000-BB000, and 8031 I/O address 63D4-63DC becomes 3D4-3DC
for the 72C81. Thus, to display an "A" in the first location of the screen, the 8031 writes the
byte 65 to location 8000. Video memory is 16K bytes of dynamic RAM (U34, U35). The

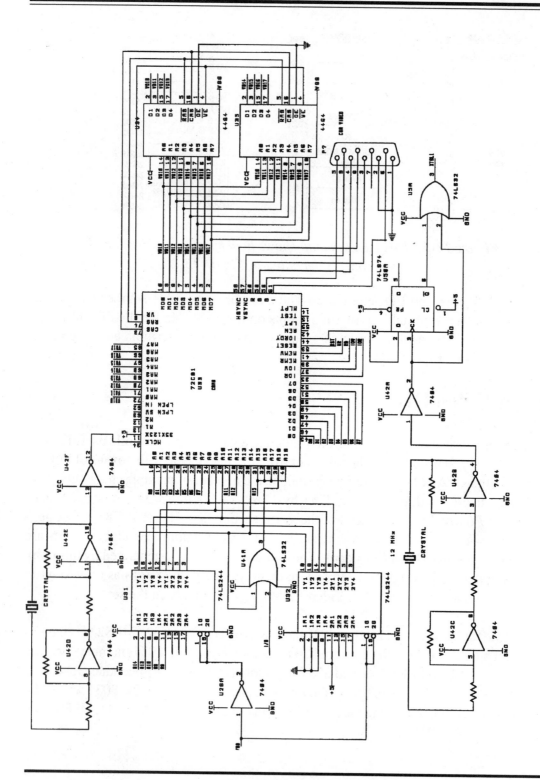

Figure 13-3. The NCR 72C81 provides a complete video interface that can be added to almost any microcontroller application.

refresh of dynamic RAM is handled by the 72C81, and it is invisible to the 8031. This screen memory is like any other memory to the 8031—it can be written to and read from.

Processors often have a *wait* or *ready* input so you can synchronize from slow external chips. The 72C81 was designed to work with such a line. The only problem is, the 8031 does not have a *wait* or *ready* input. To solve this, first use an 80C31 as the microcontroller. This CMOS version of the 8031 is fully static. This means that if you stop the clock, it retains all of its internal registers. Then instead of attaching a crystal directly to the clock input of the 80C31, you generate an external clock (U42). This clock is passed through an OR gate (U5) to the 80C31 clock input. The IORDY line from the 72C81 can turn the clock on/off (U50) to the 80C31. This simulates a *wait* function.

C Programming

Okay, the 8031 example is definitely complicated. How about the firmware? If the firmware was written in Assembly, then it would indeed be a very massive undertaking. For such a large application, a lot of time can be saved by using a C compiler. We looked briefly at C before, but for such a small example it did not show its true power. Let's examine some C code written to support the hardware design in *Figure 13-1* and *Figure 13-2*. This code was written to be compiled by the Franklin C compiler for the 8031. (With slight modifications, any 8031 C compiler should work.) The code shown is only about five percent of the total code for this particular 8031 application:

```
#define V1IREG XBYTE [0x63D4]      /* video 72C81 address register */
#define V1DREG XBYTE [0x63D5]      /* video 72C81 data register    */
#define V1MODE XBYTE [0x63D8]      /* video mode register   */
#define V1COLOR XBYTE [0x63D9]     /* video color register */
#define V1BUFF 0x8000              /* video frame buffer    */
#define S1CNTR XBYTE [0x6003]      /* S1 control register   */
#define S1BUF XBYTE [0x6000]       /* S1 rec/xt buffer      */
#define S1INT XBYTE [0x6001]       /* S1 int register       */
#define S1RINT XBYTE [0x6002]      /* S1 read int register */
#define S1MCR XBYTE [0x6004]       /* S1 modem control reg  */
#define S1STAT XBYTE [0x6005]      /* S1 status register    */
#define S1MSR XBYTE [0x6006]       /* S1 modem status reg   */
```

Just as in the previous assembler examples, you can define constants for the rest of the program. Here the address locations are defined for video I/O registers, video memory, and serial port number one. These addresses are the firmware result of the hardware address decoding in *Figure 13-1*:

```
scrnprt()                                        /* PRINT CHAR TO SCREEN */
{
unsigned short realpos;
        realpos = ((2 * pos) + 0x8000);    ;Video memory location
        XBYTE [realpos] = let;             ;store character to memory
        realpos++;
        XBYTE [realpos] = atur;            ;store attribute of character
        curright();
}

    printf(s)                              /* simple string print function */
    char *s;                               /* example call: printf("Hello, world!")
    {
            char *p = s;
        movcur();
            while(*p != '\0')
                {
                let = *p;
                scrnprt();
                p++;
                }
    }
```

These routines show how you can send any character combination desired to the screen. By issuing the command *printf("This is a test sentence which will appear on video display")* from the main program, the overhead for the video display is done.   Notice that in the subroutine *scrnprt()*, a byte called *atur* is also stored. This is set elsewhere in the program and is the attribute setting for the character. Depending on its value, a character in the text string could be displayed in color or with other special effects:

```
sendchar()
{
```

```
unsigned char *pa, a;
    if (ps1outcur != ps1outpos)
        {
        if (S1XMTHAND == 3)
            {
            a = S1MSR;
            if ((a & 16) == 16) return;
            }
        if (xt1flag == 1) return;
        pa = (pbuf1out + ps1outpos++);
        do { a = S1STAT; } while (a < 64);
        S1BUF = *pa;
        }
}
```

The routine *sendchar()* takes one character from a buffer and sends it to the serial port. Notice the use of the defined address constants to access the serial port:

```
com2int() interrupt 2 using 2
{
unsigned char *pa;
unsigned char a;
    a = S2STAT;
    if ((a & 1) > 0)
        {
        a = S2BUF;
        pa = (keybuf + keybufcur++);
        *pa = a;
        return;
        }
    a = S3STAT;
    if ((a & 1) > 0)
        {
        a = S3BUF;
        pa = (keybuf + keybufcur++);
```

```
                    *pa = a;
                    return;
                    }

        }
```

The routine *com2int()* is an interrupt-handling subroutine. When the external interrupt *int1* is activated, the program will branch to this location. Recall from *Figure 13-1* that two devices were ORed into this one interrupt. The lines $a = S2STAT$ and $a = S3STAT$ examine the status of each device to see which one sent the interrupt. The routine then services the chip and returns from interrupt.

68HC11

First introduced by Motorola in 1985, the 68HC11 has become the flagship of Motorola's 8-bit microcontroller line. Most versions of this controller contain 2 serial ports, an 8-bit A/D, timer/counter, watchdog timer, and nonvolatile EEPROM. The 68HC11 has two 8-bit accumulators which can function as one 16-bit accumulator—very useful when doing certain math routines. The current maximum clock rate is 16 MHz, which provides an internal rate of 4 MHz and approximate MIPS of 2.

The 68HC11 has many options, including D/A, PWM, a math coprocessor, and a memory-expansion unit which can increase addressing to one megabyte. The 68HC11 can have two serial ports—one for standard serial communications, and one for *serial peripheral interface* (SPI). SPI is used to expand I/O ports available to the microcontroller. One interesting ability that some 68HC11s have is converting I/O lines to programmable chip selects lines. This helps to eliminate the address decoding discussed in the pervious section.

Let's look at the features of one of the more advanced versions of the 68HC11: the 68HC11C0. This chip has 256 bytes RAM, 1K bytes ROM, and 64K bytes of external address space expandable to 256K bytes by using the built-in memory mapping logic. It also has five external chip select lines, each with programmable clock-stretching for slow devices. The 68HC11C0 has nine external interrupt request lines, one of which is NMI. Also on board is a 4-channel A/D converter. It has 35 I/O lines (31 bidirectional, 4-input only), and the bidirectional lines have programmable internal pull-up resisters. Of course, it also has all the regular features of the core 68HC11. It is hard to imagine what this chip can't do.

Scenix SX

An exciting new microcontroller is the SX18AC/SX28AC manufactured by Scenix Semiconductor. Scenix has taken the best features of the Microchip PIC line and improved it several ways. The SX is pin-and-code compatible with PIC 16C5x chips. While it is also RISC architecture, an improved instruction pipeline allows the controller to execute one instruction per clock cycle. With a maximum clock speed of 50 MHz, the SX has a whopping 50 MIPS. This is ten times the speed of comparable PIC chips.

The program memory is two thousand bytes of EEPROM. This is very convenient as all chips are reprogrammable, and there is no need to purchase a separate EPROM version of the chip for development purposes. Additionally, boards can be designed to allow updating of the controller code without removing the SX chip. There are 136 bytes of RAM memory, including both special function registers and general purpose memory.

There are several other improvements which makes this microcontroller very attractive. The stack is eight levels deep, providing for more subroutine calls. There is a single level interrupt stack, whereas the PIC 16C5x chips have none. All pins have selectable internal pullup resistors, eliminating external parts. For projects not requiring maximum speed, the SX has an internal RC oscillator which runs at 4 MHz. If an internal voltage regulator had been added, many applications could have been completed with just a battery and the SX chip!

There is one other novel feature of the SX chip. Made possible by its extreme speed, Scenix has coined the term *virtual peripherals* as a method of adding what would be normally hardware peripherals. To clarify this, consider the SPI bus mentioned in earlier chapters. With most microcontrollers, if you wanted the SPI bus, then you would need to purchase the controller version with the option. The controller itself would have more internal transistors to support this function. With the SX, to get an SPI bus would only require loading the proper software module along with your code. The chip remains the same. Of course, this occupies some of the controller processing time, but with 50 MIPS, you have plenty to spare. Here are some of the many virtual peripherals available:

- DTMF processing (reading and creating telephone tones)
- Music and voice synthesis
- PWM output
- Spectrum analysis
- Video controller

- Serial, parallel, I^2C bus, SPI bus
- IR receiver/transmitter
- Stepper-motor
- LCD module

MCS-96

The MCS-96 was one of the first 16-bit microcontrollers on the market. It was the sequel to Intel's enormously popular 8051 controller. Originally created as a custom microcontroller for automotive applications, the MCS-96 is now used for motor control, and industrial and process applications. With a maximum external clock of 20 MHz, its internal speed is 10 MHz, producing approximately 2.2 MIPS.

At this point, you may be asking, what difference does a 16-bit controller make? At 2.2 MIPS, it doesn't seem any faster than an 8-bit controller. Recall that MIPS stands for *million instructions per second*. It doesn't say anything about the *power* of those instructions. 16-bit controllers have elaborate and sophisticated instruction sets that allow actions that would take an 8-bit controller several times longer to execute. An example would be *multiply* and *divide*; the MCS-96 can execute those commands in 1.4 µs and 2.4 µs, respectively. Also, an instruction can have 1, 2 or 3 operands doing in one line of code what would take an 8-bit controller 3 lines of code to accomplish.

Standard features for the MCS-96 include 232 general-purpose registers, maximum 1K byte of RAM, 32K bytes of ROM (external 1 Mbyte addressing range), 2 timers, watchdog timer, DMA controller, serial port, maximum 56 I/O lines, and interrupts. Depending on the version, optional features include an 8-input A/D, a slave port for parallel communications with other controllers, PWM, and autoprogramming of the external EPROM. Some powerful firmware features are indirect-auto-increment addressing, table-indirect jump, and single-instruction memory block moves.

One of the most recent members of the MCS-96 family is the 8xC196. Its internal speed is a blistering 50 MHz, producing an approximate MIPS of 11. What makes this so interesting is that the internal speed, unlike other controllers we have seen, is above the external oscillator! This is possible because the 8xC196 is equipped with an on-board phase lock loop, which doubles or quadruples the external clock.

i960

With Intel's i960 microcontroller, we take the leap into 32-bit processing. This controller is a cross between a RISC and CISC. (See the previous discussion on controller architectures.) The way information flows internally makes it a RISC, and the 184-command instruction set makes it a CISC. Calculating MIPS is difficult on this type of processor, but it would be fair to say that the i960 can do an average of 150 MIPS at its high external clock speed of 100 MHz. That's right — the number of instructions executed per second is faster than the clock speed! This is because of the superscaler internal processing. Plus, each instruction is very powerful.

If you are familiar with minicomputers or mainframes, some features of the i960 will surprise you. This microcontroller has up to 4K bytes of instruction cache, a register cache which holds 15 sets of local registers, I/O and memory DMA, and 4 gigabytes of external addressing. The superscaler operation is a 4-instruction decoder which handles up to four instructions per clock cycle. This, along with all the caches, is responsible for its amazing speed.

The i960 is obviously for heavy-duty applications. You can find this microcontroller in graphic processing, telecommunications, avionics, and laser printers. The Joint Integrated Avionics Working Group has selected the i960 as the standard CPU for military flight control systems.

The Next Step

The last few sections in this book discussed advanced microcontrollers. As microcontrollers grow larger, faster and more powerful, they increasingly become similar to microprocessors. We lightly touched upon external I/O and memory chips within the architecture of a common bus (microprocessor) design. It would double the size of this book to go into all of the specifics of advanced microcontroller design, plus demonstrate working examples with projects. The next step is the study of 16-bit and 32-bit controllers, along with sophisticated 8-bit controller applications.

On the horizon and approaching fast is the fuzzy logic controller. Because these controllers are heavily analog-oriented, and the programming is much different from standard microcontrollers, we would include these chips in an advanced course on controllers. Two interesting fuzzy logic controllers are the Adaptive Logic AL220 and the SGS-Thomson W.A.R.P. 1.1. The AL220 is a small 18-pin chip which has four A/D inputs and four D/A outputs. Based upon a 10-MHz clock, the AL220 can read the inputs, evaluate "rules." and update the outputs at a constant 10,000 times per second. The W.A.R.P. (Weight Associative Rule Processor) is a much more advanced fuzzy controller with 8 inputs, 10 outputs and a maximum clock rate of 40 MHz.

To continue your studies, seek out books on these subjects. A possible sequel to this book might be forthcoming and would cover all of these areas in detail. It would also include an interesting experiment where we have a common objective for an application. The application would be designed three times with 8-bit, 16-bit and fuzzy logic microcontrollers. Prototypes would be constructed, then we would compare cost, speed and effectiveness of each design.

Until then, there are several good magazines that will teach you more about microcontrollers and update you on the latest controller developments:

EDN
EDN Circulation Manager
44 Cook Street
Denver, CO 80206
303-388-4511

Electronic Design
Penton Publishing Inc.
1100 Superior Ave.
Cleveland, OH 44114
216-696-7000

Circuit Cellar Ink
The Computer Application Journal
PO Box 698
Holmes, PA 19043
800-269-6301

Embedded Systems Programming
PO Box 420046
Palm Coast, FL 32142
800-829-5537

With the proper knowledge, virtually any electronic device can be made better with microcontrollers. It is now time for *you* to design your own microcontroller application!

APPENDIX I
Firmware Listing

Chapter 7

```
;*******************************
;* PIC PK Tester V2.0    *
;*  C. 1995              *
;*  Larry Duarte         *
;*******************************
;
;
;
; Variables and equates
;
strobe          =       ra.0
random          =       ra.1
reset           =       ra.2
statest         =       ra.3

led1            =       rb
led2            =       rc

                org     8                       ;start address of register variable space

count_low       ds      1
count_hi        ds      1
statdir         ds      1                       ;left = 0, right = 1
stat_low        ds      1
stat_hi         ds      1
led1buf ds      1
led2buf ds      1
strobecnt       ds      1
counter1        ds      1
counter2        ds      1
bitbuf          ds      1
bitcnt          ds      1
randomout       ds      1
savebit ds      1

                org     0                       ;reset code origin
```

```
;
; Device data                         CLOCK RATE 78 us
;
            device  pic16c55,rc_osc,wdt_off,protect_off
;
waitnochk                             ;time delay without
       mov    count_low,#0    ;checking for button push
       mov    count_hi,#150
nochklp
       djnz   count_low,nochklp
       djnz   count_hi,nochklp
       ret
;
wait                                  ;time delay with checking
       mov    count_low,#0    ;for buttons
       mov    count_hi,#10
waitlp
       csae   ra,#12                   ;ck for any buttons pushed
       ret
       djnz   count_low,waitlp
       djnz   count_hi,waitlp
       ret
;
test                                  ;test for randomness
       mov    counter1,#2             ;lets do it two times
test1
       mov    counter2,#240   ;2 x 240 = 480 samples
test2
       mov    led1,counter2
       mov    led2,counter2
       call   getbit
       call   statupdate
       djnz   counter2,test2
       djnz   counter1,test1           ;done 2 times?
       cja    stat_low,#48,error            ;exceeds norm by 48, test failed!
       mov    led1,#0                  ;test ok
       mov    led2,#0
       call   waitnochk
       call   resetnow
       ret
;
showstat                              ;display bias, if any
       cjb    statdir,#2,readyset
       mov    led1,#254                ;no bias
       mov    led2,#254
```

```
            call       waitnochk
            ret
readyset
            cje        stat_hi,#0,belowmax
            mov        counter2,#0                         ;max bias
            jmp        setstat
belowmax
            cja        stat_low,#32,lev2                   ;find level by grouping into 7 ranges
            mov        counter2,#254
            jmp        setstat
lev2
            cja        stat_low,#64,lev3
            mov        counter2,#252
            jmp        setstat
lev3
            cja        stat_low,#96,lev4
            mov        counter2,#248
            jmp        setstat
lev4
            cja        stat_low,#128,lev5
            mov        counter2,#240
            jmp        setstat
lev5
            cja        stat_low,#160,lev6
            mov        counter2,#224
            jmp        setstat
lev6
            cja        stat_low,#192,lev7
            mov        counter2,#192
            jmp        setstat
lev7
            mov        counter2,#128
setstat                           ;left or right?
            cje        statdir,#0,setleft
            mov        led1,#255
            mov        led2,counter2
            call       waitnochk
            ret
setsleft
            mov        led2,#255
            mov        led1,counter2
            call       waitnochk
            ret
;
statupdate                                                 ;update movement counter
```

```
        cje     statdir,#2,nodir
        cje     statdir,randomout,samedir
        djnz    stat_low,statdone               ;different dir
        cje     stat_hi,#0,setnodir
        dec     stat_hi
        ret
samedir
        ijnz    stat_low,statdone
        inc     stat_hi
        ret
nodir
        inc     stat_low
        mov     statdir,randomout
        ret
setnodir
        mov     statdir,#2
statdone
        ret
;
resetnow                                        ;reset movement counter, LEDs
        mov     led1,#254                       ;top led on
        mov     led1buf,#254
        mov     led2,#255                       ;all leds off
        mov     led2buf,#255
        mov     statdir,#2                      ;no dir
        mov     stat_low,#0
        mov     stat_hi,#0
        ret
;
getbit                                          ;random bit
        mov     bitbuf,#0
tryagain
        mov     strobecnt,#10                   ;if strobe is high 10 times, get bit  - was 8
teststrobe
        jnb     strobe,tryagain
        djnz    strobecnt,teststrobe
        movb    bitbuf.0,random                 ;get random bit
        mov     randomout,#1
        snb     bitbuf.0
        ret
        mov     randomout,#0
        ret
;
; Reset address
;
```

```
        reset   start                   ;program start
;
start                                   ;HARDWARE INITIALIZATION
        clr     fsr
        mov     !ra,#15                 ;set RA, all input
        mov     !rb,#0                  ;set RB, all output
        mov     !rc,#0                  ;set RC, all output
        movb    savebit.0,statest
        mov     counter1,#0
timeloop                                ;led test
        mov     led1,counter1
        mov     led2,#255
        call    waitnochk
        mov     led2,counter1
        mov     led1,#255
        call    waitnochk
        mov     led1,#254               ;top led on
        mov     led1buf,#254            ;VARIABLE INITIALIZATION
        mov     led2,#255               ;all leds off
        mov     led2buf,#255
        mov     statdir,#2              ;left - 0
        mov     stat_low,#0             ;right - 1
        mov     stat_hi,#0
        jb      savebit.0,loop          ;CK FOR TEST BUTTON AT START
        call    test
loop
        jb      reset,ck2               ;IS RESEST BUTTON DOWN?
        call    resetnow
        call    waitnochk
ck2
        jb      statest,donechk         ;IS STATUS BUTTON DOWN?
        call    showstat
donechk
        call    getbit                  ;GET RANDOM BIT
        call    statupdate              ;UPDATE STATUS COUNTER
        jb      bitbuf.0,highin
        csne    led1buf,#127            ;if bottom led is on, jmp to routine
        jmp     lowest
        csne    led2buf,#254            ;if top led is on, jmp
        jmp     highest
        setb    c
        rr      led2buf
        mov     led2,led2buf
        setb    c
        rl      led1buf
```

```
          mov     led1,led1buf
          jmp     delay
lowest
          mov     led1,#255
          mov     led1buf,#255
          mov     led2,#127
          mov     led2buf,#127
          jmp     delay
highest
          mov     led1,#254
          mov     led1buf,#254
          mov     led2,#255
          mov     led2buf,#255
          jmp     delay
highin
          csne    led1buf,#254          ;if top led is on, jmp
          jmp     high2
          csne    led2buf,#127          ;if bottom led is on, jmp
          jmp     low2
          setb    c
          rr      led1buf
          mov     led1,led1buf
          setb    c
          rl      led2buf
          mov     led2,led2buf
          jmp     delay
low2
          mov     led1,#127
          mov     led1buf,#127
          mov     led2,#255
          mov     led2buf,#255
          jmp     delay
high2
          mov     led1,#255
          mov     led1buf,#255
          mov     led2,#254
          mov     led2buf,#254
delay
          call    wait                  ;SPEED DELAY
          jmp     loop
error                                   ;ERROR, endless loop
          mov led1,#0
          mov led2,#0
          call waitnochk
          mov led1,#255
```

```
        mov led2,#255
        call waitnochk
        jmp error
;
;end
;
```

Chapter 8

```
;***************************
;
;*  RS232 Terminal      *
;*  C. 1996             *
;*  Larry Duarte        *
;***************************
;
;
;
; Variables and equates
;
rs232_out     =      ra.2
rs232_in      =      ra.3

keyb          =      rb                  ;keyboard

lcd_reg       =      ra.1                ;hi = data register, low  = instruction register
lcd_clk       =      ra.0                ;normal low (tri-state), data moves on falling edge

lcd           =      rc

              org    8                   ;start address of file register variable space

count_low     ds     1                   ;8
count_hi      ds     1
out_char      ds     1
temp          ds     1
in_char       ds     1
flag          ds     1                   ;13
char_count    ds     1
char_loc      ds     1
key_temp      ds     1
char          ds     1
key_in        ds     1
old_key       ds     1                   ;19
```

```
flag_in          =        flag.0                  ;high if char has been rec
flag_out         =        flag.1                  ;high if char is to be sent rs232
flag_shift       =        flag.2                  ;high if shift is pressed
flag_key         =        flag.3                  ;high if we have a key

                 org    0                         ;reset code origin
;
; Device data
;
                          device  pic16c55,hs_osc,wdt_off,protect_off
;
wait                                              ; .217us clock
        djnz count_low,wait                       ; full count = .04277 sec
        djnz count_hi,wait
        ret
;
lcd_wait
        clr       count_low
        mov       count_hi,#1
lcdwait_loop
        djnz      count_low,lcdwait_loop
        djnz      count_hi,lcdwait_loop
        ret
;
wait_check
        clr       count_low
        mov       count_hi,#50
ck_loop
        jnb       rs232_in,ck_exit   ;if incoming on rs232, exit now!
        djnz      count_low,ck_loop
        djnz      count_hi,ck_loop
ck_exit
        ret
;
one_bit                                           ;send one bit
        rr        out_char                        ;set up carry with bit to send
        sc
        clrb      rs232_out
        snc
        setb      rs232_out
        mov       count_low,#152
        mov       count_hi,#1
        ret
;
get_bit                                           ;pull in one bit
```

```
        sb      rs232_in                    ;test rs232, set up carry for rotate
        clc
        snb     rs232_in
        stc
        rr      in_char
        mov     count_low,#154
        mov     count_hi,#1
        ret
;
rec_char                                    ;get one char from rs232
        mov     count_low,#75
        mov     count_hi,#1
        call    wait
        snb     rs232_in
        ret                                 ;false start, get out of here
        mov     temp,#8                     ;OK get eight bits
        mov     count_low,#154
        mov     count_hi,#1
rec_loop
        call    wait
        call    get_bit
        djnz    temp,rec_loop
        call    wait
        setb    flag_in                     ;rec char ready
        ret
;
send_char
        clrb    rs232_out                   ;start bit
        mov     count_low,#158
        mov     count_hi,#1
        call    wait

        mov     temp,#8 ;send eight bits
send_loop
        call    one_bit                     ;send single bit
        call    wait
        djnz    temp,send_loop

        setb    rs232_out                   ;stop bit
        mov     count_low,#158
        mov     count_hi,#1
        call    wait

        ret
;
```

```
trans_key
        csne    key_in,#119
        retw    48                          ;return from table with 1 in work register
        csne    key_in,#183
        retw    49
        csne    key_in,#215
        retw    50
        csne    key_in,#231
        retw    51
        csne    key_in,#123
        retw    52
        csne    key_in,#187
        retw    53
        csne    key_in,#219
        retw    54
        csne    key_in,#235
        retw    55
        csne    key_in,#125
        retw    56
        csne    key_in,#189
        retw    57
        csne    key_in,#221
        retw    65
        csne    key_in,#237
        retw    66
        csne    key_in,#126
        retw    67
        csne    key_in,#190
        retw    68
        csne    key_in,#222
        retw    69
        csne    key_in,#238
        retw    70
        retw    0                           ;default return
;
clr_screen                                  ;clear LCD screen
        mov     lcd,#1
        call    lcd_wait
        setb    lcd_clk
        call    lcd_wait
        clrb    lcd_clk
        call    lcd_wait
        clr     char_count
        ret
;
```

```
prt_char
        mov     lcd,char                        ;setup data lines for lcd
        call    lcd_wait
        setb    lcd_reg                         ;line high for data
        call    lcd_wait
        setb    lcd_clk
        call    lcd_wait
        clrb    lcd_clk                         ;strobe clock
        call    lcd_wait
        clrb    lcd_reg
        call    lcd_wait
        inc     char_count
        ret
prt_loop
        cjne    char_count,#8,prt_char          ;need to change screen?
        mov     lcd,#168                        ;change address for part II of screen
        setb    lcd_clk
        call    lcd_wait
        clrb    lcd_clk
        call    lcd_wait
        jmp     prt_char                        ;do char now


;
; Reset address
;
        org     256                             ;page 1 code
        reset   start
;
start
        clr     fsr
        mov     !ra,#8                          ;set IO lines port A
        mov     ra,#0
        mov     !rb,#240                            ;set IO lines port B 11110000
        mov     !rc,#0                          ;set IO lines port C all out
        setb    rs232_out                       ;rs232 is normal high
        clr     flag                            ;clr all flags
        clr     old_key
        mov     key_in,#255                     ;no key
        clrb    lcd_reg                         ;init LCD
        clrb    lcd_clk
        mov     lcd,#48
        setb    lcd_clk
        call    lcd_wait
        clrb    lcd_clk
        call    lcd_wait
```

```
        setb    lcd_clk
        call    lcd_wait
        clrb    lcd_clk
        call    lcd_wait
        setb    lcd_clk
        call    lcd_wait
        clrb    lcd_clk                      ;send #48 three times
        call    lcd_wait
        mov     lcd,#56
        setb    lcd_clk
        call    lcd_wait
        clrb    lcd_clk
        call    lcd_wait
        mov     lcd,#15
        setb    lcd_clk
        call    lcd_wait
        clrb    lcd_clk
        call    lcd_wait
        mov     lcd,#1
        setb    lcd_clk
        call    lcd_wait
        clrb    lcd_clk
        call    lcd_wait
        mov     lcd,#6
        setb    lcd_clk
        call    lcd_wait
        clrb    lcd_clk
        call    lcd_wait
        mov     count_hi,#255
        clr     char_count
        call    wait
loop
        cje     key_in,#255,nokey            ;key to send?
        mov     out_char,key_in
        call    send_char
nokey
        call    wait_check                   ;delay for debounce, chk rs232
        jb      rs232_in,no_in               ;see if incoming on rs232
        call    rec_char
        cjne    in_char,#13,no_cr            ;carriage return?
        call    clr_screen                   ;yes, clear screen
        jmp     no_in
no_cr
        cjb     char_count,#16,ok_toprt        ;16 chars?
        call    clr_screen                   ;auto return after 16 chars
```

```
        call    wait_check
ok_toprt
        mov     char,in_char
        call    prt_loop
no_in
        mov     key_in,#255
        mov     keyb,#14
        mov     char,keyb               ;read keyboard
        cjb     char,#240,got_key
        mov     keyb,#13
        mov     char,keyb
        cjb     char,#240,got_key
        mov     keyb,#11
        mov     char,keyb
        cjb     char,#240,got_key
        mov     keyb,#7
        mov     char,keyb
        cjb     char,#240,got_key
        clr     old_key                 ;no key pressed
        jmp     loop
got_key
        cje     char,old_key,noneed     ;same key pressed
        mov     old_key,char            ;save old scan code
        mov     key_in,char
        call    trans_key               ;translate key
        mov     key_in,w
noneed
        jmp     loop
;
;end
;
```

Chapter 9

```
;****************************
;
;*  ACMOTOR.SRC      *
;*  Larry A. Duarte  *
;*  C. 1996          *
;*  AC Motor Control *
;****************************
;
;
; Device data
;
```

```
                        device  pic16c54,rc_osc,wdt_on,protect_off
;
;
;
; Variables and equates
;
mc1                 =       ra.0            ;mc1 - mc4 control triacs
mc2                 =       ra.1
mc3                 =       ra.2
mc4                 =       ra.3            ;all out - control register 0

aux                 =       rb.0
but_up              =       rb.1
but_down            =       rb.3
but_stop            =       rb.2
limit_top           =       rb.4
limit_bottom        =       rb.5
drum                =       rb.6
led                 =       rb.7            ;control register 01111110 - 126

                    org     7               ;start address of file register variable space

count_low           ds      1               ;low byte delay counter
count_hi            ds      1               ;high byte delay counter
num_blink           ds      1               ;number of blinks for led
temp                ds      1
count               ds      1
count2              ds      1
drum_count_low              ds      1               ;low byte drum count
drum_count_hi       ds      1               ;high byte drum count
error_code          ds      1               ;number of LED blinks for error
flag                ds      1

flag_slot_state     =       flag.0          ;current state of drum sensor
flag_pause          =       flag.1          ;pause current opperation
flag_direction      =       flag.2          ;direction of lift

motor_stop          =       0
motor_up            =       14
motor_down          =       13

set_low             =       50              ;distance of motor travel
set_hi              =       25              ;Example: each notch on slot wheel is
                                            ;equal to 1/10" of linear travel
                                            ;total distance = ((25 * 255) + 50) * 1/10"
                                            ;           = 53 1/2 feet
```

```
stuck_slot       =        60                        ;count down for stuck slot

                 org      0                          ;reset code origin
;
wait                                                 ; 38us clock, 152us per instruction
        djnz     count_low,wait                      ; 3 cycles + 3 more evey 256
        djnz     count_hi,wait                       ; full count = 30 seconds
        ret
;
ledloop                                              ;make LED blink number of times in num_blink
        clr      wdt                                 ;clear watchdog timer
        setb     led
        clr      count_low
        mov      count_hi,#2
        call     wait
        clrb     led
        clr      count_low
        mov      count_hi,#2
        call     wait
        djnz     num_blink,ledloop
        ret


start_down
        jnb      limit_bottom,down_exit              ;at bottom limit, return
        cjb      drum_count_hi,#set_hi,down_ok;chk set limit
        cja      drum_count_hi,#set_hi,down_exit     ;way too low, return
        cjb      drum_count_low,#set_low,down_ok     ;chk set limit
        ret                                          ;all ready at set position
down_ok
        jnb      flag_pause,down_now                 ;not moving
        jnb      flag_direction,down_exit            ;all ready moving down?
        call     start_stop                          ;must be moving up, stop it
        clr      count_low
        mov      count_hi,#8                         ;wait 1 seconds
        clr      wdt                                 ;clear watchdog timer
        call     wait
down_now
        setb     flag_pause                          ;moving
        clrb     flag_direction                      ;downward
        mov      count2,#stuck_slot*2                ;stuck drum counter
        movb     flag_slot_state,drum                ;save drum state
        mov      ra,#motor_down
down_exit
        ret
;
```

```
start_up
        sb      limit_top                       ;are we at top?
        ret                                     ;if yes then return
        jnb     flag_pause,up_now               ;not moving
        jb      flag_direction,up_exit          ;all ready moving up?
        call    start_stop                      ;must be moving down, stop it
        clr     count_low
        mov     count_hi,#8                     ;wait 1 seconds
        clr     wdt                             ;clear wdt
        call    wait
up_now
        setb    flag_pause
        setb    flag_direction
        mov     count2,#stuck_slot*2            ;we load twice normal stuck
        movb    flag_slot_state,drum            ;slot count because motor
        mov     ra,#motor_up                    ;take a while to get up to
up_exit                                         ;speed
        ret


start_stop
        jnb     flag_pause,stop_exit            ;all ready stopped
        mov     ra,#motor_stop
        clrb    flag_pause
        clr     count_low                       ;must build little timer to avoid sub-call
        mov     count_hi,#2                     ;little less than 1 second
little_timer
        clr     wdt                             ;clear watchdog timer
        jb      flag_slot_state,stop_low_state  ;look at old state
        jnb     drum,stop_dec_count             ;is drum low, if so jump
        jmp     stop_do_math
stop_low_state
        jb      drum,stop_dec_count             ;is drum hi, if so jump
stop_do_math
        movb    flag_slot_state,drum            ;save drum state
        jb      flag_direction,stop_up_math     ;add or subtract
        ijnz    drum_count_low,stop_dec_count       ;increment and jump if not zero
        inc     drum_count_hi
        jmp     stop_dec_count
stop_up_math
        dec     drum_count_low                      ;going up, decrement counter
        cjne    drum_count_low,#255,stop_dec_count
        dec     drum_count_hi
stop_dec_count
        djnz    count_low,little_timer
        djnz    count_hi,little_timer
```

```
stop_exit
        ret
;
;
; Reset address
;
        org     256
        reset   start
;
start
        mov     ra,#motor_stop              ;make sure tiracs are off
        mov     !ra,#0                      ;set ra to all out
        mov     ra,#motor_stop              ;extra sure triacs are off
        mov     !rb,#126                         ;set IO lines port b
        clr     wdt                         ;clear watchdog timer
        clrb    flag_pause                  ;we are not moving
        movb    flag_slot_state,drum        ;save sensor state
        mov     num_blink,#2                ;2 blinks
        call    ledloop                     ;blink hello
        clr     drum_count_low
        clr     drum_count_hi
main_loop
        lset    main_loop                   ;reset page
        clr     wdt                         ;clear watchdog timer
        mov     !ra,#0                      ;reset IO direction ra, rb
        mov     !rb,#126                         ;in case of hiccup
ml_ck_buts
        sb      but_down                    ;is down pressed?
        call    start_down                  ;process down button
        sb      but_up                      ;is up pressed?
        call    start_up                    ;process up button
        sb      but_stop                    ;is stop pressed?
        call    start_stop
        jnb     flag_pause,main_loop        ;we are not moving
ml_ck_sensor
        djnz    count2,ml_not_stuck         ;is drum moving?
        mov     ra,#motor_stop              ;STUCK!
        mov     error_code,#4
        jmp     error_end
ml_not_stuck
        jnb     flag_direction,ml_ck_down   ;ck for limits & position - up or down
        jb      limit_top,ml_drum_slot      ;jmp if not up
        call    start_stop                  ;we are up
        clr     drum_count_low
        clr     drum_count_hi               ;reset drum counter
```

```
           jmp     main_loop                        ;stop
ml_ck_down
           jnb     limit_bottom,down_stop
           cjb     drum_count_hi,#set_hi,ml_drum_slot     ;chk set limit
           cjb     drum_count_low,#set_low,ml_drum_slot ;chk set limit
down_stop
           call    start_stop                       ;we are at set position
           jmp     main_loop
ml_drum_slot
           jb      flag_slot_state,ml_low_state     ;look at old state
           jnb     drum,main_loop                   ;is drum low, if so jump
           jmp     ml_do_math
ml_low_state
           jb      drum,main_loop                   ;is drum hi, if so jump
ml_do_math
           movb    flag_slot_state,drum             ;save drum state
           mov     count2,#stuck_slot               ;detected movement, reset stuck counter
           jb      flag_direction,ml_up_math        ;add or subtract
           ijnz    drum_count_low,main_loop         ;increment and jump if not zero
           inc     drum_count_hi
           jmp     main_loop
ml_up_math
           dec     drum_count_low                            ;going up, decrement counter
           cjne    drum_count_low,#255,main_loop
           dec     drum_count_hi
           jmp     main_loop
;
error_end
           clr     wdt                              ;clear watchdog timer
           mov     ra,#motor_stop
           mov     !ra,#0
           mov     ra,#motor_stop
           mov     !rb,#126
           mov     num_blink,error_code
           call    ledloop
           clr     count_low
           mov     count_hi,#10
           call    wait
           jmp     error_end                        ;endless loop
;end
;
```

Chapter 10

```
;********************************
;
;*  Busy Buster             *
;*  C. 1996                 *
;*  Larry Duarte            *
;********************************
;
;
; Device data           Clock Rate 108 us
;
                device  PIC16C54,rc_osc,wdt_off,protect_off

;
; Variables and equates
;
spk             =       ra.0                    ;speaker
hook            =       ra.1                    ;on-off hook switch
rso             =       ra.2                    ;register select line M-8888

rd              =       rb.4                    ;read line M-8888
wr              =       rb.5                    ;write line M-8888
cs              =       rb.6                    ;chip select line M-8888
irq             =       rb.7                    ;IRQ/CP line M-8888

                org     7                       ;start address of file register variable space

temp1           ds      1
data            ds      1                       ;data to send/receive from M-8888
count_low       ds      1
counthi         ds      1
pnlen           ds      1                       ;lenght of number dialed
pn1             ds      1                       ;phone number - register 12
pn2             ds      1
pn3             ds      1
pn4             ds      1
pn5             ds      1
pn6             ds      1
pn7             ds      1
pn8             ds      1
pn9             ds      1
pn10            ds      1
pn11            ds      1                       ;phone number - register 22
                                                ;room for ten more numbers!

                org     0                       ;reset code origin
```

```
;
getnum
        setb    rso                             ;stat reg
        call    r8888                           ;read status
        clrb    rso                             ;back to data
        call    r8888                           ;read 8888
        mov     temp1,data
        and     temp1,#1111b                    ;mask off junk
        mov     indirect,temp1                  ;store into num string
        inc     fsr                             ;inc indirect
        ret
;
dial
        mov     fsr,#12                         ;prepare FSR for indirect addressing
dial2
        mov     data,indirect
        call    w8888
        inc     fsr
        csb     fsr,pnlen
        jmp     dialdone
        mov     counthi,#2                      ;164ms
        mov     count_low,#250
        call    wait
        jmp     dial2
dialdone
        setb    rso
        call    r8888                           ;clr interupt
        clrb    rso
        ret
;
w8888                                           ;write to M-8888
        mov     !rb,#128  ;data out
        setb    data.4    ;rd
        setb    data.5    ;wr
        clrb    data.6    ;cs
        mov     rb,data
        clrb    wr
        setb    wr
        setb    cs
        mov     !rb,#143
        ret
;
r8888                                           ;read from M-8888
        clrb    cs
        clrb    rd
```

```
          mov     data,rb
          setb    rd
          setb    cs
          ret
;
beep                                          ;beep from speaker
          setb    spk
          mov     count_low,#250
belp
          clrb    spk                         ;square wave to speaker
          nop
          setb    spk
          djnz    count_low,belp
          clrb    spk
          mov     counthi,#1
          mov     count_low,#250
          call    wait
          ret
;
wait
          djnz    count_low,wait              ; .324ms * (count_low - 1)
          djnz    counthi,wait                ;plus 83ms * (counthi - 1)
          ret                                 ;
;
; Reset address
;
          reset   start
;
start
          mov     !ra,#8                      ;set RA
          mov     ra,#0
          mov     !rb,#143                         ;set RB
          mov     rb,#255
          call    beep                        ;two beeps for power ok
          call    beep
          mov     data,#13                    ;tone on,irq on, reg b next
          setb    rso                         ;control reg
          call    w8888
          mov     data,#0                     ;burst mode, no test
          call    w8888
          clrb    rso                         ;back to data mode
          mov     w,#7                        ;7: rtcc = .108us * 256 = 27.64ms, .02764 sec
          option                              ;store into option register
          mov     fsr,#12                     ;prepare FSR for indirect addressing
          call    getnum                      ;misc startup number, clear M-8888 buffer
```

```
        mov     fsr,#12                 ;reset address
stchk                                   ;loop here for phone number
        csb     rb,#128                 ;more numbers?
        jmp     stchk                   ;no
        call    getnum                  ;yes
        csb     temp1,#11               ;if * or # goto gotnum
        jmp     gotnum                  ;loop back
        jmp     stchk
gotnum
        dec     fsr                     ;get rid of *
        mov     pnlen,fsr               ;we got number and busy
        call    beep
        call    beep
        call    beep
        mov     counthi,#48             ; 4 seconds
        mov     count_low,#250
        call    wait                    ;wait for them to hang up phone
        setb    rso
        mov     data,#7                  ;set 8888 for cp,burst mode, tone on
        call    w8888
        clrb    rso
mkcall
        clrb    hook                    ;hangup
        call    beep                    ;dial beep
        mov     counthi,#12
        mov     count_low,#250          ;1 sec wait on hangup
        call    wait
        setb    hook                    ;off hook
        mov     counthi,#12             ;
        mov     count_low,#250          ;must wait 1 sec to dial
        call    wait
        call    dial
cklowset
        mov     temp1,ra                ;make sure det is low to start
        and     temp1,#1000b
        csb     temp1,#1
        jmp     cklowset                ;its high so goback
detlow
        mov     temp1,ra
        and     temp1,#1000b            ;only det is left
        csb     temp1,#1                ;checking for high or low on det
        jmp     dethigh                 ;det is high
        jmp     detlow
dethigh
        mov     rtcc,#0
```

```
dethigh2
        mov     temp1,ra
        and     temp1,#1000b            ;only det is left
        csb     temp1,#1                ;checking for high or low on det
        jmp     dethigh2                ;det is high
        mov     temp1,rtcc              ;det is low, look at time
        csb     temp1,#14               ;less than .4 sec?
        jmp     cktme                   ;no
        jmp     detlow                  ;back to start
cktme
        csb     temp1,#25               ;less than .7 sec
        jmp     aok                     ;no, ringing
        jmp     mkcall                  ;we got busy
aok
        call    beep                    ;endless beeps
        jmp     aok

;
;end
;
```

Chapter 11

```
;********************
;
;*  Video.src      *
;*  C. 1996        *
;*  Larry Duarte   *
;********************
;
;
        device pic16c71,hs_osc,wdt_off,pwrt_off,protect_off
;
;       variables and equates
;
blank         = ra.0                   ;drops video line to .25 volts
sync_io       = ra.2                   ;drops video line to 0 volts
temp1_data    = rb.0                   ;temp sensor 1
temp1_clk     = rb.1
temp1_cs      = rb.2
temp2_data    = rb.3                   ;temp sensor 2
temp2_clk     = rb.4
temp2_cs      = rb.5


sync          = 104                    ;Load into RTCC with prescalar of 2 for 63.6 us
```

```
vert_time1  =   119                          ;bigger, the shorter time
temp1_in    =   9                            ;00001001    Change IO direction between read
temp1_out   =   8                            ;00001000    and write
temp2_in    =   9                            ;00001001
temp2_out   =   1                            ;00000001

            org         12

flag            ds      1                    ;flags
vert_count      ds      1                    ;number of vertical pulses
hor_count       ds      1                    ;number of blank lines
line_count      ds      1                    ;number of data lines
w_copy          ds      1                    ;storage area for w register
s_copy          ds      1                    ;storage area for status register
count5          ds      1                    ;counter used in 5us delay
row             ds      1                    ;current row of line
col_count       ds      1                    ;current collum of line
temp_temp       ds      1                    ;temperature register
temp            ds      1                    ;temporary math register
count           ds      1                    ;
count_low       ds      1                    ;
in_temp         ds      4                    ;inside temp
out_temp        ds      4                    ;outside temp
screen_buf      ds      7                    ;pixel bits for one horzional line
line_buf        ds      7                    ;characters for current line


flag_vert       =       flag.0               ;currently in vertical retrace portion
flag_hor        =       flag.1               ;currently doing 15 blank lines
flag_line       =       flag.2               ;curretnly doing 245 video lines
flag_go         =       flag.3               ;do we have info to show?
flag_sign       =       flag.4               ;plus or neg C

            org     0
            jmp     start            ;PIC16C71 allways start at location 0, jmp to real start

            org     4                        ;interrupt address
            jmp     handler
;
get_matrix                                   ;get pixel pattern for character
            add     w,row        ;enter with character number in w (ex: 1 is 12, 2 is 24)
            jmp     pc+w         ;then add current screen row for pixels
            retw    0c3h,81h,00,18h,18h,3ch,3ch,18h,18h,00,81h,0c3h      ;O
            retw    0c3h,0c3h,0e3h,0e3h,0e3h,0e3h,0e3h,0e3h,0e3h,0e3h,00,00   ;1
            retw    0c3h,81h,18h,18h,0f1h,0c3h,87h,8fh,1fh,1fh,00,00    ;2
```

```
        retw    0c3h,81h,18h,0f8h,0f8h,0c1h,0c1h,0f8h,0f8h,18h,81h,0c3h        ;3
        retw    18h,18h,18h,18h,00,00,00,0f8h,0f8h,0f8h,0f8h,0f8h       ;4
        retw    00,00,1fh,0fh,87h,0c3h,0e1h,0f0h,0f8h,18h,81h,0c3h      ;5
        retw    0c3h,81h,18h,18h,1fh,07h,03h,11h,18h,18h,81h,0c3h       ;6
        retw    00,00,00,0f0h,0f0h,0f0h,0f0h,0f0h,0f0h,0f0h,0f0h,0f0h    ;7
        retw    0c3h,81h,18h,18h,81h,0c3h,0c3h,81h,18h,18h,81h,0c3h ;8
        retw    0c3h,81h,18h,18h,88h,0c0h,0e0h,0f8h,18h,18h,81h,0c3h       ;9
        retw    00,00,0c3h,0c3h,0c3h,0c3h,0c3h,0c3h,0c3h,0c3h,00,00 ;I
        retw    18h,18h,08h,00,00,00,10h,18h,18h,18h,18h,18h           ;N
        retw    0c3h,81h,18h,18h,8fh,0c3h,0e1h,0f1h,18h,18h,81h,0c3h ;S
        retw    07h,03h,01h,10h,18h,18h,18h,18h,10h,01h,03h,07h        ;D
        retw    00,00,1fh,1fh,1fh,00,00,1fh,1fh,1fh,00,00               ;E
        retw    0ffh,0ffh,0ffh,0ffh,0ffh,0ffh,0ffh,0ffh,0ffh,0ffh,0ffh,0ffh  ;
        retw    18h,18h,18h,18h,18h,18h,18h,18h,18h,00,81h,0c3h       ;U
        retw    00h,00h,00h,0c3h,0c3h,0c3h,0c3h,0c3h,0c3h,0c3h,0c3h,0c3h  ;T
        retw    0ffh,0ffh,0e7h,0e7h,0e7h,081h,081h,0e7h,0e7h,0e7h,0ffh,0ffh  ;+
        retw    0ffh,0ffh,0ffh,0ffh,0ffh,081h,081h,0ffh,0ffh,0ffh,0ffh,0ffh  ;-
;
handler                                             ;interrupt routine
        mov     w_copy,w                            ;save copy of work register
        mov     s_copy,status                       ;save copy of status register
        clrb    T0IF                                ;clear timer interrupt flag
        mov     TMR0,#sync                          ;ready for next interrupt
        jb      flag_line,video_line                ;are we doing video line?
        jb      flag_hor,blank_line                 ;are we doing blank lines?
                                                    ;must be in vertical retrace period

        cjavert_count,#0,next_vert
        setb    sync_io
        incvert_count
        mov     TMR0,#vert_time1                    ;because the vertical pulses are inverted
        jmp     h_exit                              ;from normal sync, interrupt time is different
next_vert
        clrb    sync_io
        call    wait_5us
        setb    sync_io
        incvert_count
        cjevert_count,#4,vert_done                  ;after three vertical pulses
        mov     TMR0,#vert_time1
        jmp     h_exit
vert_done
        clrb    flag_vert                           ;vertical retrace pulse done
        setb    flag_hor                            ;prepare for blank lines
        clr     hor_count                           ;clear counter
        jmp     blank_line
;
```

```
video_line
        setb     sync_io                          ;normal video line
        call     wait_5us                         ;start with horizonal sync
        clrb     sync_io

        incline_count                            ;increment line count
        cjeline_count,#240,line_done             ;are we done with frame?
do_char
        jnb      flag_go,char_exit                ;is this a blank video line?
pre_load
        mov      w,line_buf                       ;the time between the end of
        call     get_matrix                       ;sync and the start of the
        mov      screen_buf,w                     ;real display is used to pre-load
        mov      w,line_buf+1                     ;the screen buffer with pixels
        call     get_matrix                       ;this provides for max speed
        mov      screen_buf+1,w                   ;during the display
        mov      w,line_buf+2                      ;this could have been done with
        call     get_matrix                       ;a loop but did not do so because
        mov      screen_buf+2,w                    ;of the extra time overhead
        mov      w,line_buf+3
        call     get_matrix
        mov      screen_buf+3,w
        mov      w,line_buf+4
        call     get_matrix
        mov      screen_buf+4,w
        mov      w,line_buf+5
        call     get_matrix
        mov      screen_buf+5,w
        mov      w,line_buf+6
        call     get_matrix
        mov      screen_buf+6,w

        mov      fsr,#screen_buf                  ;prepare for indirect addressing of screen buffer
        clrb     blank                            ;line before character for box
        mov      col_count,#7                     ;number of bytes in screen buffer
        nop                                       ;need for width of line
        setb     blank                            ;end line before character
next_char                                         ;we will now create pixels on screen
        rl       indirect                         ;indirect is pointing to pixel screen buffer
        rl       ra                               ;rotate left buffer, pixel goes to carry bit
        rl       indirect        ;rotate left I/O port, carry sets high or low on video line
        rl       ra                               ;this way a pixel is set in two instructions
        rl       indirect                         ;or 0.4 us
        rl       ra                               ;this is also why we grounded pin RA1
        rl       indirect                         ;to prevent rotating a high into the sync
```

```
        rl      ra                      ;pin RA2
        rl      indirect                ;no loop on this code to save time
        rl      ra
        rl      indirect
        rl      ra
        rl      indirect
        rl      ra
        rl      indirect
        rl      ra
        incfsr                          ;next screen buffer
        setb    blank                      ;black after character
        djnz    col_count,next_char        ;time for space between characters
        clrb    blank                      ;line after character for box
        incrow
        nop
        nop                             ;time for line width
        setb    blank                   ;end box line
        cjerow,#12,end_line         ;end of one character line?
        jmp     h_exit
char_exit                                  ;on blank lines we use the time
        cjeline_count,#60,set_line1    ;to set the character buffer
        cjeline_count,#80,set_line2
        cjeline_count,#120,set_line4
        cjeline_count,#140,set_line5
        cjaline_count,#161,h_exit      ;totaly blank, not even doing box
        cjbline_count,#53,h_exit       ;totaly blank, not even doing box
        jmp     do_box                     ;do box
line_done
        clrb    flag_line                  ;character line done
        setb    flag_vert                  ;clear flags and counter
        clr     vert_count
        jmp     h_exit
;
blank_line                                 ;blank lines
        setb    sync_io                    ;simulates 'equalizing' pulses
        call    wait_5us
        clrb    sync_io
        inchor_count
        cjehor_count,#20,hor_done      ;19 pulses then done
        jmp     h_exit
hor_done
        clrb    flag_hor                   ;clear flag and counter
        setb    flag_line
        clr     line_count
;
```

```
h_exit
        mov     status,s_copy                ;restore status register upon exit
        swap    w_copy                       ;from interrupt
        mov     w,<>w_copy                   ;swap work register into w, status is unaffected
        reti
;
start                                        ;program start
;
temp_config                                  ;set DS1620 config
        mov     !rb,#temp1_out
        clrb    temp1_cs
        setb    temp1_clk
        setb    temp1_cs
        clrb    temp1_clk
        mov     temp_temp,#0Ch               ;set DS1620 to proper mode
        mov     count,#8
temp_wl1
        clrb    temp1_clk
        rr      temp_temp                    ;put bit into carry
        movb    temp1_data,c                 ;put carry onto data line
        setb    temp1_clk
        djnz    count,temp_wl1
        mov     temp_temp,#2                 ;config byte
        mov     count,#8
temp_wl2
        clrb    temp1_clk
        rr      temp_temp                    ;put bit into carry
        movb    temp1_data,c                 ;put carry onto data line
        setb    temp1_clk
        djnz    count,temp_wl2
        clrb    temp1_cs
        mov     !rb,#temp1_in
;
temp2_config                                 ;set temp config for 2nd sensor
        mov     !rb,#temp2_out
        clrb    temp2_cs
        setb    temp2_clk
        setb    temp2_cs
        clrb    temp2_clk
        mov     temp_temp,#0Ch               ;put DS1620 into proper mode
        mov     count,#8
temp2_wl1
        clrb    temp2_clk
        rr      temp_temp                    ;put bit into carry
        movb    temp2_data,c                 ;put carry onto data line
```

```
        setb    temp2_clk
        djnz    count,temp2_wl1
        mov     temp_temp,#2                    ;config byte
        mov     count,#8
temp2_wl2
        clrb    temp2_clk
        rr      temp_temp                       ;put bit into carry
        movb    temp2_data,c                    ;put carry onto data line
        setb    temp2_clk
        djnz    count,temp2_wl2
        clrb    temp2_cs
        mov     !rb,#temp2_in
;
        call    one_sec                         ;time after write function on DS1620
;
start_temp                                      ;must now tell DS1620 to go
        mov     !rb,#temp1_out
        clrb    temp1_cs
        setb    temp1_clk
        setb    temp1_cs
        clrb    temp1_clk
        mov     temp_temp,#0EEh                 ;start conversions
        mov     count,#8
temp_st1
        clrb    temp1_clk
        rr      temp_temp
        movb    temp1_data,c
        setb    temp1_clk
        djnz    count,temp_st1
        mov     !rb,#temp1_in
        clrb    temp1_cs
;
start2_temp                                     ;same for 2nd sensor
        mov     !rb,#temp2_out
        clrb    temp2_cs
        setb    temp2_clk
        setb    temp2_cs
        clrb    temp2_clk
        mov     temp_temp,#0EEh                 ;start conversions
        mov     count,#8
temp2_st1
        clrb    temp2_clk
        rr      temp_temp
        movb    temp2_data,c
        setb    temp2_clk
```

```
        djnz      count,temp2_st1
        mov       !rb,#temp2_in
        clrb      temp2_cs
;                                         ;prepare PIC registers
        setb      rp0                     ;switch to register page 1
        clr       wdt                     ;assign prescaler to RTCC
        mov       option,#0               ;prescaler divide by 2
        mov       adcon1,#3               ;configure port A as non-A/D
        clrb      rp0                     ;switch to register page 0
        mov       TMR0,#sync              ;start sync time delay
        mov       intcon,#10100000b       ;interrupt register set for only RTCC interrupts
        mov       !ra,#0                  ;port A all output
        clrb      sync_io                 ;no sync
        setb      blank                   ;video blank level
        clr       flag                    ;clear all flags
        clr       vert_count              ;clear counters
        clr       hor_count
        clr       line_count
        clr       row
        setb      flag_vert               ;start on vertical pulse
; init temp registers                    ;dummy values for temp numbers
        mov       in_temp,#216            ;for split second at start
        clr       in_temp+1
        clr       in_temp+2
        clr       in_temp+3
        mov       out_temp,#216
        clr       out_temp+1
        clr       out_temp+2
        clr       out_temp+3
;
main                                      ;program loop
;
read_temp                                 ;read sensor #1
        mov       !rb,#temp1_out
        clrb      temp1_cs
        setb      temp1_clk
        setb      temp1_cs
        clrb      temp1_clk
        mov       temp_temp,#0AAh         ;read temp command
        mov       count,#8
temp_wl3
        clrb      temp1_clk
        rr        temp_temp               ;put bit into carry
        movb      temp1_data,c            ;put carry onto data line
        setb      temp1_clk
```

```
        djnz       count,temp_wl3
;
        mov        !rb,#temp1_in                      ;get data
        mov        count,#8                           ;8 data bits
temp_rd1
        clrb       temp1_clk
        movb       c,temp1_data
        rr         temp_temp
        setb       temp1_clk
        djnz       count,temp_rd1
;
        clrb       temp1_clk                          ;9th bit is sign
        movb       c,temp1_data                       ;+ or - bit
        setb       temp1_clk
        clrb       temp1_cs
;
        call       translate                          ;convert from C to F
;
        jb         flag_sign,temp1_neg                ; sign bit
        mov        in_temp,#216
        jmp        digit1
temp1_neg
        mov        in_temp,#228
;                  translate from binary to 3 digit number
digit1
        clr        count
d1_sub
        cjb        temp_temp,#100,d1_done
        add        count,#12                          ;one 100, index by 12 (see matrix table)
        sub        temp_temp,#100
        jmp        d1_sub
d1_done
        mov        in_temp+1,count                    ;save in in_temp
digit2
        clr        count
d2_sub
        cjb        temp_temp,#10,d2_done
        add        count,#12                          ;one 10, index by 12  (see matrix table)
        sub        temp_temp,#10
        jmp        d2_sub
d2_done
        mov        in_temp+2,count                    ;save in in_temp
digit3
        clr        count
        cje        temp_temp,#0,d3_done
```

```
d3_sub
     add       count,#12                        ;one's, index by 12
     djnz      temp_temp,d3_sub
d3_done
     mov       in_temp+3,count
;
read2_temp                                      ;read 2nd sensor
     mov       !rb,#temp2_out
     clrb      temp2_cs
     setb      temp2_clk
     setb      temp2_cs
     clrb      temp2_clk
     mov       temp_temp,#0AAh                  ;read temp command
     mov       count,#8
temp2_wl3
     clrb      temp2_clk
     rr        temp_temp                        ;put bit into carry
     movb      temp2_data,c                     ;put carry onto data line
     setb      temp2_clk
     djnz      count,temp2_wl3
;
     mov       !rb,#temp2_in                    ;get data
     mov       count,#8
temp2_rd1
     clrb      temp2_clk
     movb      c,temp2_data
     rr        temp_temp
     setb      temp2_clk
     djnz      count,temp2_rd1
;
     clrb      temp2_clk
     movb      c,temp2_data                     ;+ or - bit
     setb      temp2_clk
     clrb      temp2_cs
;
     call      translate                        ;convert from C to F
;
     jb        flag_sign,temp2_neg              ; sign bit
     mov       out_temp,#216
     jmp       odigit1
temp2_neg
     mov       out_temp,#228
;                translate from binary to 3 digit number
odigit1
     clr       count
```

```
od1_sub
    cjb         temp_temp,#100,od1_done
    add         count,#12                       ;one 100, index by 12
    sub         temp_temp,#100
    jmp         od1_sub
od1_done
    mov         out_temp+1,count                ;save in out_temp
odigit2
    clr         count
od2_sub
    cjb         temp_temp,#10,od2_done
    add         count,#12                       ;one 10, index by 12
    sub         temp_temp,#10
    jmp         od2_sub
od2_done
    mov         out_temp+2,count                ;save in out_temp
odigit3
    clr         count
    cje         temp_temp,#0,od3_done
od3_sub
    add         count,#12                       ;one's, index by 12
    djnz        temp_temp,od3_sub
od3_done
    mov         out_temp+3,count
;
    jmp main                                    ;end main program loop
;
wait_5us
    mov         count5,#5                       ;5us time delay
loop5
    djnz        count5,loop5
    ret
;
wait                                            ;small, tight variable time delay
    djnz        count_low,wait
    ret
;
one_sec                                         ;about 39 ms
    clr         count_low
    clr         count
sec_loop
    djnz        count_low,sec_loop
    djnz        count,sec_loop
    ret
;
```

```
translate                                    ;change from centigrade to fahrenheit
        movb    flag_sign,c                  ;(C x 1.8) + 32 = F
        clr     count                        ;the number from DS1620 is 1/2 degree C per bit
        jb      flag_sign,neg_temp
        mov     temp,temp_temp               ;store into temp
        jmp     trans_loop
neg_temp                                     ;temp is negitive
        clr     temp_temp
        sub     temp_temp,temp               ;0 - temp
        mov     temp,temp_temp               ;store into temp register as above
trans_loop
        cjb     temp,#10,no_more
        sub     temp,#10
        inc     count
        jmp     trans_loop
no_more
        sub     temp_temp,count
        jnb     flag_sign,add32              ;sign is pos
        cja     temp_temp,#32,f_neg          ;sign will be neg
        mov     temp,#32
        sub     temp,temp_temp
        clrb    flag_sign
        mov     temp_temp,temp
        ret
f_neg                                        ;the fahrenheit reading is negitive
        sub     temp_temp,#32
        ret
add32
        add     temp_temp,#32
        ret
;
end_line                                     ;end character line
        clrb    flag_go                      ;clear flags and counter
        clr     row
        jmp     h_exit
;
set_line1                                    ;set character buffer 'INSIDE'
        setb    flag_go
        mov     line_buf,#120
        mov     line_buf+1,#132
        mov     line_buf+2,#144
        mov     line_buf+3,#120
        mov     line_buf+4,#156
        mov     line_buf+5,#168
        mov     line_buf+6,#180
```

```
        jmp         box2a
;
set_line2                                       ;set character buffer with sensor #1 temp
        setb        flag_go
        mov         line_buf,#180
        mov         line_buf+1,in_temp
        mov         line_buf+2,in_temp+1
        mov         line_buf+3,in_temp+2
        mov         line_buf+4,in_temp+3
        mov         line_buf+5,#180
        mov         line_buf+6,#180
        jmp         box2b
;
set_line4                                       ;set character buffer with 'OUTSIDE'
        setb        flag_go
        mov         line_buf,#0
        mov         line_buf+1,#192
        mov         line_buf+2,#204
        mov         line_buf+3,#144
        mov         line_buf+4,#120
        mov         line_buf+5,#156
        mov         line_buf+6,#168
        jmp         box2c
;
set_line5                                       ;set character buffer with sensor #2 temp
        setb        flag_go
        mov         line_buf,#180
        mov         line_buf+1,out_temp
        mov         line_buf+2,out_temp+1
        mov         line_buf+3,out_temp+2
        mov         line_buf+4,out_temp+3
        mov         line_buf+5,#180
        mov         line_buf+6,#180
        jmp         box2d
;
do_box                                          ;parts of screen box
        cjb         line_count,#55,box1a
        cje         line_count,#105,box1b
        cje         line_count,#106,box1c
        cja         line_count,#159,box1
        jmp         box2
box2a                                           ;nop's are used to even out the timing
        nop
        nop
        nop
```

```
        nop
box2b
        nop
        nop
        nop
        nop
box2c
        nop
        nop
        nop
        nop
box2d
        mov     count_low,#1
        call    wait
        nop
        nop
box2
        mov     count_low,#5
        call    wait
        clrb    blank
        nop
        mov     count_low,#47
        setb    blank
        call    wait
        nop
        nop
        clrb    blank
        nop
        nop
        nop
        setb    blank
        jmp     h_exit
box1a
        nop
        nop
        nop
        nop
box1b
        nop
        nop
        nop
        nop
box1c
        nop
        nop
```

```
        nop
        nop
box1
        mov     count_low,#5
        call    wait
        nop
        clrb    blank
        mov     count_low,#49
        call    wait
        nop
        nop
        setb    blank
        jmp     h_exit
;
;end
;
```

APPENDIX II
Project Part List

Chapter 7

SEMICONDUCTORS

U1 — PIC® 16C55 microcontroller
U2 — LM311 comparator
U3 — 555 timer
U4 — LM317 adjustable voltage regulator
U5 — 7805 5-volt regulator
Q1, Q2 — 2N4401 general-purpose NPN transistor
D1, D2 — IN4148 small-signal diode
LED1 - LED16 — Red light-emitting diode

RESISTORS

(All fixed resistors are 1/4-watt, 5% units, unless otherwise noted)
R1, R14 — 220-ohm
R2, R3, R4, R11, R12 — 10,000-ohm
R5 — 10,000-ohm, 15-turn potentiometer
R6, R8 — 1-megohm
R7, R10 — 100,000-ohm
R9 — 100-ohm
R13 — 1,500-ohm
R15 — 47,500-ohm, 1%
R16 — 42,200-ohm, 1%

CAPACITORS

C1, C5-220-pF, ceramic-disc
C2, C4, C9 — 0.1-uF, polyester
C3 — 0.0047-uF, polyester
C6, C12 — 0.01-uf, polyester
C7 — 10-uF, 25-WVDC, electrolytic
C8 — 100-uF, 16-WVDC, electrolytic

C10 — 1000-uF, 16-WVDC, electrolytic
C11 — 0.001-uF, polyester

ADDITIONAL PARTS

J1 — Mono phone jack
S1, S2 — Normally open SPST, momentary push-button switch
S3 — SPST switch
IC sockets, LED lens caps, 12-volt AC adapter

Chapter 8

SEMICONDUCTORS

U1 — PIC16C55 microcontroller
U2 — MAX232 RS-232 interface
Q3 — 7805 5-volt regulator

RESISTORS

(All fixed resistors are 1/4-watt, 5% units, unless otherwise noted)
R1, R2, R3, R4 — 3,300-ohm
R5 — 10,000-ohm potentiometer

CAPACITORS

C1 — 100-uF, 25-WVDC, electrolytic
C2,C3 — 0.1-uF, polyester
C4, C5, C6, C7 — 1-uF, 50-WVDC, electrolytic
C8, C9 — 15-pF, ceramic-disc

ADDITIONAL PARTS

X1 — 18.43 MHz crystal
N1 — power jack
N2 — DB9 connector, PCB mount
N3 — 8 pin .1 header
N4 — 14 pin .1 header
LCD — Optrex OP116-ND
4x4 keypad, cables, 12-volt AC adapter

Chapter 9

SEMICONDUCTORS

U1 — PIC® 16C54 microcontroller
U2, U3, U4, U5, U6 — MOC3061, zero-crossing optocoupler
Q1 — 7805 5-volt regulator
D1 — Bridge
TR1, TR2, TR3, TR4, TR5 — Q4015L6, Teccor Alternistor

RESISTORS

(All fixed resistors are 1/4-watt, 5% units, unless otherwise noted)
R1, R8 — 100,000-ohm
R2, R3, R4, R6, R7, R15, R18, R21, R24, R27 — 1,000-ohm
R5, R9, R11, R12, R13, R14 — 470-ohm
R10 — 180-ohm
R17, R20, R23, R26, R29 — 24-ohm
R16, R19, R22, R25, R28 — 100-ohm

CAPACITORS

C1, C4 — 0.1-uF, polyester
C2 — 100-uF, 25-WVDC, electrolytic
C3 — 330-pF, ceramic-disc
C5, C6, C7, C8, C9 — 0.1-uF, 250-WVDC, polyester

ADDITIONAL PARTS

T1 — Power transformer, 12-volt output
N2 — 10 pin .1 header
Slot Sensor — QVB11134, Quality Technologies
3 Normally open, SPST, momentary push-button switch
2 SPST switches, AC cord with in-line fuse, Red LED, cables
Reversing three wire AC motor with brake

Chapter 10

SEMICONDUCTORS

U1 — PIC® 16C54 microcontroller
U2 — M8888 DTMF telephone interface
Q1 — ZVNL110A, N-MOSFET
Q2 — 78L05 5-volt regulator
D2, D3, D4, D5 — 1N4001 rectifier
D1 — 1N4148 diode

RESISTORS

(All fixed resistors are 1/4-watt, 5% units, unless otherwise noted)
R1 — 100,000-ohm
R2 — 6,800-ohm
R3 — 390,000-ohm
R4 — 47,000-ohm
R5 — 220,000-ohm
R6 — 680-ohm
R7 — 3,300-ohm
R8 — 680-ohm, 1/2 watt

CAPACITORS

C1 — 220-pF, ceramic-disc
C2 — 2.2-uF, 10-WVDC, electrolytic
C3, C4, C5 — 0.1-uF, polyester
C6 — 0.1-uF, 100-WVDC, polyester
C7 — 1-uF, 100-WVDC, electrolytic
C8 — 100-uF, 10-WVDC, electrolytic

ADDITIONAL PARTS

X1 — 3.579 MHz crystal
T1 — 600 to 600 ohm telephone transformer
S1 — slide switch

MOV1 — Metal Oxide Varistor, 130-volt
N4, N5 — RJ11 PCB mount jacks
Piezo buzzer

Chapter 11

SEMICONDUCTORS

U1 — PIC® 16C54 microcontroller
U2 — 7805 5-volt regulator
U3, U4 — DS1620, Digital thermometer and thermostat
Q1 — 2N222 NPN transistor
Q2 — MPS3904 NPN transistor

RESISTORS

(All fixed resistors are 1/4-watt, 5% units, unless otherwise noted)
R1, R2 — 470-ohm
R3 — 33-ohm
R4 — 220-ohm
R5 — 75-ohm

CAPACITORS

C1, C2 — 15-pF, ceramic-disc
C3 — 100-uF, 25-WVDC, electrolytic
C4 — .1-uF, polyester

ADDITIONAL PARTS

X1 — 20 MHz crystal
N1 — 10 pin .1 header
N2 — RCA-Jack PCB mount
N3 — Power jack, or battery connector
12-volt AC adapter, cables

APPENDIX III
Vendors

Company	Product
Adaptive Logic, Inc. 800 Charcot Ave. #112 San Jose, CA 95131 408-383-7201 http://www.adaptivelogic.com/	AL220 fuzzy microcontroller
Allen Systems 2346 Brandon Road Columbus, OH 43221 614-488-7122	8051 development board
Analog Devices One Technology Way Norwood, MA 02062 617-461-3881	Analog semiconductor devices, A/D, D/A, DSP
Coactive PO Box 425967 SF, CA 94142 415-626-5152	68HC11 development board
Custom Computer Services PO Box 11191 Milwaukee, WI 53211 414-781-2794	PIC® C compiler
Dallas Semiconductor 4401 S. Beltwood Parkway Dallas, TX 75244 214-450-0448	Microcontrollers Other semiconductors DS1620 (Digital thermometer and thermostat)

Digi-Key Corporation
701 Brooks Ave. South
Thief River Falls, MN 56701-0677
800-344-4539
http://www.digikey.com

Electronic components

Most parts for the projects in
this book can be acquired
from this company.

Franklin Software, Inc.
888 Saratoga Ave., #2
San Jose, CA 95129
408-296-8051

C compiler
Debugger/simulator
Emulator software

IAR Systems Software, Inc.
One Maritime Plaza
San Francisco, CA 94111
415-765-5500

C compiler for 8051, 80251,
80196, 68HC11 and others.

ImageCraft
PO Box 64226
Sunnyvale, CA 94088
408-749-0702
imagecft@netcom.com

68HC11 C compiler

IMP, Inc.
2830 N. First St.
San Jose, CA 95134-2071
408-434-1377

EPAC, analog processing

Intel Corporation
3065 Bowers Ave.
Santa Clara, CA 95051
800-628-8686

Microcontrollers
8051 plus many others

J & M Microtek, Inc.
83 Seaman Road
W. Orange NJ 07052
201-325-1892

80C552 single board
computer, universal simulator,
EPROM emulator, PIC® in-circuit
emulation/programmer

Jameco   Electronic components
1355 Shoreway Road
Blemont, CA 94002
415-592-8097

JDR Microdevices   Electronic components
1850 South 10th Street
San Jose, CA 95112
800-538-5000
http://www.jdr.com

Keil Software   8051/251 assembler, C
16990 Dallas Parkway, Suite 120   compiler
Dallas, TX 75248
800-348-8051

Micro Engineering Labs   PIC® prototyping boards
Box 7532   PIC® programmer
Colorado Springs, CO 80933
719-520-5323

Micro-Bit Technology, Inc.   Line of embedded controllers
2201 Midway Rd. Suite #106   based on the 80C51
Carrollton, TX 75006
214-386-9294

Microchip Technology, Inc.   Microcontrollers
2355 West Chandler Blvd.   PIC® family line
Chandler, AZ 85224-6199
602-963-7373

MicroMint   A whole range of
4 Park Street   microcontroller products.
Vernon, CT 06066
203-875-2751

Motorola
PO Box 20912
Phoenix, AZ 85036
602-244-6609
http://Design-NET.com

Microcontrollers
68HC11 family line

Mouser Electronics
2401 HWY 287
N. Mansfield, TX 76063-4827
800-346-6873
http://www.mouser.com

Electronic components

National Semiconductor
800-272-9959
http://www.natsemi.com

COP8 microcontroller
A/D, D/A
Other semiconductors

Nohau Corporation
3830 Del Amo Blvd., Suite #101
Torrance, CA 90503
310-214-1791

Optical electronics
LCD

Palomar Telecom, Inc.
1201 Simpson Way
Escondido, CA 92029
619-746-7998

Self-contained natural voice
playback module

Parallax, Inc.
3805 Atherton Road, #102
Rocklin, CA 95765
916-624-8333
parallaxinc.com

PIC® programmer, emulator,
simulator, assembler
BASIC stamp

Teltone Corporation
10801 120th Ave. NE
Kirkland, WA 98033
800-426-3926

Telephone chips
M8888

Toshiba America Electronic 9775 Toledo Way Irvine, CA 92718 714-453-0224	Microcontrollers 4-bit, 8-bit, and 16-bit lines
Tribal America Electronic 44388 S. Grimmer Blvd. Fremont, CA 94538 510-623-8859	Programmers In-circuit emulators
VLSI Technology 8375 River Parkway Tempe, AZ 85284 602-753-6373	ARM processor
Zeta Electronic Design 7 Colby Ct. Unit #4-193 Bedford, NH 03110 603-644-3239	PIC® prototype board
Zilog 210 East Hacienda Ave. Campbell, CA 95008-6600 408-370-8000	Microcontrollers Z8 family line

APPENDIX IV
CD-ROM Contents

The second edition of this book includes a CD-ROM. While containing the source and binary code for the examples in this book, the CD-ROM also has many microcontroller data sheets which will be invaluable to the beginning microcontroller user. While these data sheets are highly technical, the reader by completing this book should be familiar enough with microcontroller terms to be able to read and understand the sheets. Additionally, the CD-ROM has sample assemblers and other useful files.

There are eleven directories in the root of the CD-ROM. This appendix list is in each of the directories, and shows how to access them.

BOOKCODE

Directory BOOKCODE contains the code for the examples in this book, chapters seven to eleven. There are four subdirectories in BOOKCODE. The text source code for each chapter is in subdirectory SOURCE. The compiled output files ready for programming (with the Parallax programmer) are in BINARY. A listing file for each chapter is in LISTING. Listing the files shows how the compiler translated the source code into binary, and what memory was used. Finally, MICROCHP contains the binary files in a format used by most other programmers, including the Microchip programmer.

To access the source or listing files, a simple editor or word processor is sufficient. DOS EDIT works quite nicely, or a word processor such as Microsoft Word. If you do used a word processor, open and the save the file in plain text (ASCII) format. Saving a program in the native word processor format introduces special format characters that will cause a compiler to fail.

Feel free to use the code samples in any way you like. By mixing parts of the code and the appropriate editing of memory locations, dozens of new projects can be instantly created.

Acrobat

The industry standard format for technical sheet publishing is Acrobat by Adobe Systems. Adobe provides Acrobat Reader free of charge, enabling users to view documents product by Acrobat. There are two main advantages to Acrobat. The document is formatted to look exactly like what you would see in a printed data book, including illustrations and fonts. The second advantage is that there are readers available for every major operating system.

This directory contains programs for installing Acrobat Reader onto your computer. Select the appropriate file for your operating system, run or install the file, and follow the instructions on the screen. After installation, you may then used Acrobat Reader to view all the data sheets on this CD-ROM:

- Windows 95/NT - ar32e30.exe
- Windows 3.1 - ar16e30.exe
- SGI - sgiirix.tar
- SUN - solaris.tar, sunds.tar
- Macintosh - ardr30e.bin
- DOS - acrodos.exe

Microchip

All of the examples in this book were done using Microchip's line of PIC microcontrollers. The data sheets are a valuable reference for a better understanding of the projects, and are essential for the creation of new applications:

- 30234d.pdf - PIC 16C6X
- 30235f.pdf - PIC 16C62X(A)
- 30264a.pdf - PIC 17C75X
- 30390e.pdf - PIC 16C7X
- 30412c.pdf - PIC 17C4X
- 30445c.pdf - PIC 16C84
- 30453a.pdf - PIC 16C5X
- 30561a.pdf - PIC 12C67X
- 40139c.pdf - PIC 12C5XX

- 40172a.pdf - PIC 12CE5XX
- 40181a.pdf - PIC 12CE67X

SCENIX

This is a relative newcomer to the field of microcontrollers. However, the SX chip shows great promise for the future:

- sx_datasheet.pdf - SX18AC / SX28AC
- sx_prod_brief.pdf - Specifications for the SX chip.

Intel

The giant that gave us the PC also provides microcontrollers. In fact, Intel created the first microcontroller. Intel has a complete line of microcontrollers, from 8 to 64 bits:

- 27041907.pdf - 80C31BH/80C51BH/87C51
- 27062201.pdf - Application Note, Small DC motor control
- 27231802.pdf - MCS-51
- 27245905.pdf - 8XC196NP
- 27278303.pdf - 8X251SA/SB/SP/SQ
- 27281401.pdf - 8XC151SA/SB

Motorola

Another giant is Motorola, which offers a complete line of microcontrollers. Many of their microcontrollers are highly customized for specific applications:

- an1733.pdf - Application Note, Implementing Caller ID
- an463.pdf - Application Note, Infrared remote control
- b6r4.pdf - HC05 (68HC05)
- bd3r0.pdf - HC05 (68HC05BD3)
- c0r1_2.pdf - 68HC05C0
- casm05.exe - compiler
- math16a.asm - math routines for 68HC05 microcontrollers

Texas Instruments

More well-known for its calculators and TTL product line, Texas Instruments also produces an impressive line of microcontrollers:

- spna017.pdf - TMS370 Microcontroller Family Application Book
- spnd005.pdf - TMS370 Data Book
- spnu127a.pdf - TMS370 Users Guide

Zilog

Zilog's microcontrollers are very powerful, with extensive instruction sets and abundant RAM memory. Many specialized versions of their basic microcontrollers are available:

- 86c04c08.pdf - Z86C04/C08
- e303140.pdf - Z86144
- z86c61.pdf - Z86C61/62/96
- z86e04.pdf - Z86E04/E08

Harris

- fn1328.pdf - CDP6402/CDP6402C (UART)
- fn2748.pdf - CDP68HC05CD, etc.
- fn2767.pdf - CA3282 (octal power driver)
- fn3646.pdf - HIP7030A2
- fn4249.pdf - CDP68HC05C16B

Cypress

The universal serial bus (USB) is an upcoming standard among personal computers. Cypress offers a line of microcontrollers with built-in USB support:

- 7c63000.pdf - CY7C63000
- 7c63400.pdf - CY7C63410/11/12/13

- 7c66000.pdf - CY7C66011/12/13
- usbmouse.pdf - Application Note, How to design a USB mouse

Parallax

Parallax produces development tools for microcontrollers. The examples in the book were written in Parallax assembly language. Parallax has provided a number of files as samples of their products:

- spasm.exe - Compiler for PIC microcontrollers
- stamp.exe, stamp2.exe - Editors for basic stamp modules
- commands.txt, manual.txt - Documentation for basic stamp modules
- sxkey.exe - Editor / compiler for Scenix SX microcontrollers.

Index

PROMPT®
PUBLICATIONS

Computer Monitor
Troubleshooting & Repair
by Joe Desposito & Kevin Garabedian

The explosion of computer systems for homes, offices, and schools has resulted in a subsequent demand for computer monitor repair information. *Computer Monitor Troubleshooting & Repair* makes it easier for any technician, hobbyist or computer owner to successfully repair dysfunctional monitors. Learn the basics of computer monitors with chapters on tools and test equipment, monitor types, special procedures, how to find a problem and how to repair faults in the CRT. Other chapters show how to troubleshoot circuits such as power supply, high voltage, vertical, sync and video.

This book also contains six case studies which focus on a specific model of computer monitor. The problems addressed include a completely dead monitor, dysfunctional horizontal width control, bad resistors, dim display and more.

Troubleshooting & Repair
308 pages • paperback • 8-1/2 x 11"
ISBN: 0-7906-1100-7 • Sams: 61100
$29.95 • July 1997

Complete VCR
Troubleshooting & Repair
by Joe Desposito & Kevin Garabedian

Though VCRs are complex, you don't need complex tools or test equipment to repair them. *Complete VCR Troubleshooting and Repair* contains sound troubleshooting procedures beginning with an examination of the external parts of the VCR, then narrowing the view to gears, springs, pulleys, belts and other mechanical parts. This book also shows how to troubleshoot tuner/demodulator circuits, audio and video circuits, special effect circuits, sensors and switches, video heads, servo systems and many more.

Complete VCR Troubleshooting & Repair also contains nine detailed VCR case studies, each focusing on a particular model of VCR with a very specific and common problem. The case studies guide you through the repair from start to finish, using written instruction, helpful photographs and Howard W. Sams' own *VCRfacts®* schematics. Some of the problems covered include failure to rewind, tape loading problems and intermittent clock display.

Troubleshooting & Repair
184 pages • paperback • 8-1/2 x 11"
ISBN: 0-7906-1102-3 • Sams: 61102
$29.95 • March 1997

Troubleshooting & Repair Guide to TV

by The Engineers of Howard W. Sams & Co.

The Howard W. Sams Troubleshooting & Repair Guide to TV is the most complete and up-to-date television repair book available. Included in its more than 200 pages is complete repair information for all makes of TVs, timesaving features that even the pros don't know, comprehensive basic electronics information, and coverage of common TV symptoms.

This repair guide is illustrated with useful photos, schematics, graphs, and flowcharts. It covers audio, video, technician safety, test equipment, power supplies, picture-in-picture, and much more. *The Howard W. Sams Troubleshooting & Repair Guide to TV* was written, illustrated, and assembled by the engineers and technicians of Howard W. Sams & Company. This book is the first truly comprehensive television repair guide published in the 90s, and it contains vast amounts of information never printed in book form before.

In-Home VCR Mechanical Repair & Cleaning Guide

by Curt Reeder

Like any machine that is used in the home or office, a VCR requires minimal service to keep it functioning well and for a long time. However, a technical or electrical engineering degree is not required to begin regular maintenance on a VCR. *The In-Home VCR Mechanical Repair & Cleaning Guide* shows readers the tricks and secrets of VCR maintenance using just a few small hand tools, such as tweezers and a power screwdriver.

This book is also geared toward entrepreneurs who may consider starting a new VCR service business of their own. The vast information contained in this guide gives a firm foundation on which to create a personal niche in this unique service business.

This book is compiled from the most frequent VCR malfunctions author Curt Reeder has encountered in the six years he has operated his in-home VCR repair and cleaning service.

Troubleshooting & Repair
238 pages • paperback • 8-1/2 x 11"
ISBN: 0-7906-1077-9 • Sams: 61077
$29.95 • June 1996

Troubleshooting & Repair
222 pages • paperback • 8-3/8 x 10-7/8"
ISBN: 0-7906-1076-0 • Sams: 61076
$19.95 • April 1996

CALL 1-800-428-7267 TODAY FOR THE NAME OF YOUR NEAREST PROMPT PUBLICATIONS DISTRIBUTOR

PROMPT
PUBLICATIONS

Build Your Own Home Lab
by Clement Pepper

Tube Substitution Handbook
by William Smith & Barry Buchanan

For electronic enthusiasts, a home lab is a place where learning can take place. All of the lab details – work space, equipment, parts, library – are necessary for a working facility. *Build Your Own Home Lab* shows you how to assemble an efficient working home lab. Learn how to create an ideal lab inexpensively, and how to make it pay its own way through years of growth and use. This book includes many projects for creating your own instruments, such as a multichannel oscilloscope switch and a 100-minute timer/stop watch.

The following information is covered: work space furnishings, basic tools and instrumentation, assembling a working library, making your own instrumentation, breadboarding and prototype modules, circuit data basics and much, much more! Also included are lists of mail and telephone order sources and semiconductor literature sources.

Tube substitution is one of the only feasible methods to repair or restore original tube equipment, but it should not be performed haphazardly. This handbook, itemizing all known vacuum tubes that have been or are still being manufactured, along with their replacements, will make tube substitution not only possible but relatively straightforward and efficient. The most accurate, up-to-date guide available, the *Tube Substitution Handbook* is useful to antique radio buffs, classic car enthusiasts, ham operators, and collectors of vintage ham radio equipment. In addition, marine operators, microwave repair technicians, and TV and radio technicians will find the *Handbook* to be an invaluable reference tool. The *Tube Substitution Handbook* is divided into three sections, each preceded by specific instructions. These sections are vacuum tubes, picture tubes, and tube basing diagrams.

Professional Reference
314 pages • paperback • 7-3/8 x 9-1/4"
ISBN: 0-7906-1108-2 • Sams: 61108
$24.95 • July 1997

Professional Reference
149 pages • paperback • 6 x 9"
ISBN: 0-7906-1036-1 • Sams: 61036
$16.95 • March 1995

CALL 1-800-428-7267 TODAY FOR THE NAME OF YOUR NEAREST PROMPT PUBLICATIONS DISTRIBUTOR

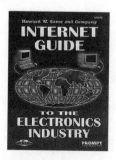

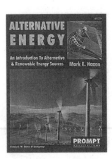

PUBLICATIONS

Disk Included!

Disk Included!

PC Hardware Projects, Volume 1
by James "J.J." Barbarello

PC Hardware Projects, Volume 2
by James "J.J." Barbarello

As an electronics hobbyist and PC owner, you probably own a digital multimeter and a logic probe. These inexpensive test and measurement instruments provide enough test muscle for many tasks, but just don't have enough capability for serious work in the world of digital electronics.

Using commonly available components and standard construction techniques, *PC Hardware Projects* will guide you through the construction of an 8-In/8-Out channel logic analyzer and a 25-line multipath continuity tester. Once you are done, you will be able to put them to immediate use with the full-featured software included with the book. You will also be able to combine the 8/8CLA, the MCT25, appropriate power supply, and a prototyping solderless breadboard system into a single digital workstation interface. *PC Hardware Projects* guides you through every step of the construction process and shows you how to check your progress.

If you have always wanted to work with stepper motors and need a rig to drill your own PC boards quickly and accurately, get out your soldering iron and follow along. The second volume in a two-part series, *PC Hardware Projects* discusses stepper motors, how they differ from conventional and servo motors, and how to control them. It investigates different methods to control stepper motors, and provides you with circuitry for a dedicated IC controller and a component hardware controller.

Then *PC Hardware Projects* guides you through every step of the construction process of an automated, PC-controlled PCB drilling machine which can drill as many as 500 holes in a 6 x 8" PCB with excellent repeatability and resolution. With construction completed, you'll then walk through an actual design layout, creating a PC design and board. You'll see how the drill data file is determined from the layout, and drill the PCB.

Projects
256 pages • paperback • 7-3/8 x 9-1/4"
ISBN: 0-7906-1104-X • Sams: 61104
$24.95 • February 1997

Projects
191 pages • paperback • 7-3/8 x 9-1/4"
ISBN: 0-7906-1109-0 • Sams: 61109
$24.95 • May 1997

CALL 1-800-428-7267 TODAY FOR THE NAME OF YOUR NEAREST PROMPT PUBLICATIONS DISTRIBUTOR

PC Hardware Projects, Volume 3
by James "J.J." Barbarello

PC Hardware Projects, Volume 3 shows you how to construct ComponentLAB, a device that connects to the digital I/O card of previous volumes and gives you the capability to measure and record DC voltage, resistance and capacitance, and test digital ICs. ComponentLAB also contains an 8-bit A/D converter that you can use to capture any analog signal ranging from a few hundred millivolts all the way up to 5 volts. This book also explains how to export the data you store into other applications for your own data processing needs, and provides you with all the inside information on how to use the digital I/O card for other purposes.

PC Hardware Projects, Volume 3 comes with a disk containing the application programs for ComponentLAB, the A/D converter, and the digital I/O card. It also contains definition files for 95 of the most common TTL and CMOS ICs, and various sample data files.

Projects
204 pages • paperback • 7-3/8 x 9-1/4"
ISBN: 0-7906-1151-1 • Sams: 61151
$24.95 • February 1998

Real-World Interfacing
With Your PC
Second Edition
by James "J.J." Barbarello

As the computer becomes increasingly prevalent in society, its functions and applications continue to expand. Modern software allows users to do everything from balance a checkbook to create a family tree. Interfacing, however, is truly the wave of the future for those who want to use their computer for things other than manipulating text, data, and graphics.

Real-World Interfacing With Your PC provides all the information necessary to use a PC's parallel port as a gateway to electronic interfacing. In addition to hardware fundamentals, this book provides a basic understanding of how to write software to control hardware. While the book is geared toward electronics hobbyists, it includes a chapter on project design and construction techniques, a checklist for easy reference, and a recommended inventory of starter electronic parts to which readers at every level can relate.

Projects
120 pages • paperback • 7-3/8 x 9-1/4"
ISBN: 0-7906-1145-7 • Sams: 61145
$24.95 • October 1997

PROMPT®
PUBLICATIONS

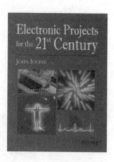

Electronic Projects for the 21st Century
by John Iovine

If you are an electronics hobbyist with an interest in science, or are fascinated by the technologies of the future, you'll find *Electronic Projects for the 21ˢᵗ Century* a welcome addition to your electronics library. It's filled with nearly two dozen fun and useful electronics projects designed to let you use and experiment with the latest innovations in science and technology. This book contains the expert, hands-on guidance and detailed instructions you need to perform experiments that involve genetics, lasers, holography, Kirlian photography, alternative energy sources and more. You will obtain all the information necessary to create the following: biofeedback/lie detector device, ELF monitor, Geiger counter, MHD generator, expansion cloud chamber, air pollution monitor, laser power supply for holography, pinhole camera, synthetic fuel from coal, and much more.

RadioScience Observing
Volume 1
by Joseph Carr

Among the hottest topics right now are those related to radio: radio astronomy, amateur radio, propagation studies, whistler and spheric hunting, searching for solar flares using Very Low Frequency (VLF) radio and related subjects. Author Joseph Carr lists all of these under the term "radioscience observing" — a term he has coined to cover the entire field.

In this book you will find chapters on all of these topics and more. The main focus of the book is for the amateur scientist who has a special interest in radio. It is also designed to appeal to amateur radio enthusiasts, shortwave listeners, scanner band receiver owners and other radio hobbyists.

RadioScience Observing also comes with a CD-ROM containing numerous examples of radio frequencies so you can learn to identify them. It also contains detailed information about the sun, planets and other planetary bodies.

Projects
256 pages • paperback • 7-3/8 x 9-1/4"
ISBN: 0-7906-1103-1 • Sams: 61103
$19.95 • June 1997

Communications Technology
336 pages • paperback • 7-3/8 x 9-1/4"
ISBN: 0-7906-1127-9 • Sams: 61127
$29.95 • January 1998

CALL 1-800-428-7267 TODAY FOR THE NAME OF YOUR NEAREST PROMPT PUBLICATIONS DISTRIBUTOR

Desktop Digital Video
by Ron Grebler

The Video Hacker's Handbook
by Carl Bergquist

Desktop Digital Video is for those people who have a good understanding of personal computers and want to learn how video (and digital video) fits into the bigger picture. This book will introduce you to the essentials of video engineering, and to the intricacies and intimacies of digital technology. It examines the hardware involved, then explores the variety of different software applications and how to utilize them practically. Best of all, *Desktop Digital Video* will guide you through the development of your own customized digital video system. Topics covered include the video signal, digital video theory, digital video editing programs, hardware, digital video software and much more.

Geared toward electronic hobbyists and technicians interested in experimenting with the multiple facets of video technology, *The Video Hacker's Handbook* features projects never seen before in book form. Video theory and project information is presented in a practical and easy-to-understand fashion, allowing you to not only learn how video technology came to be so important in today's world, but also how to incorporate this knowledge into projects of your own design. In addition to the hands-on construction projects, the text covers existing video devices useful in this area of technology plus a little history surrounding television and video relay systems.

Video Technology
225 pages • paperback • 7-3/8 x 9-1/4"
ISBN: 0-7906-1095-7 • Sams: 61095
$29.95 • June 1997

Video Technology
336 pages • paperback • 7-3/8 x 9-1/4"
ISBN: 0-7906-1126-0 • Sams: 61126
$24.95 • September 1997

AGREEMENT

READ THIS AGREEMENT BEFORE OPENING THE SOFTWARE PACKAGE

BY OPENING THE SEALED PACKAGE YOU ACCEPT AND AGREE TO THE FOLLOWING TERMS AND CONDITIONS PRINTED BELOW. IF YOU DO NOT AGREE, DO NOT OPEN THE PACKAGE AND RETURN THE SEALED PACKAGE AND ALL MATERIALS YOU RECEIVED TO HOWARD W. SAMS & COMPANY, 2647 WATERFRONT PARKWAY EAST DRIVE SUITE 100 INDIANAPOLIS, IN 46214-2041 (HEREINAFTER "LICENSOR") WITHIN 30 DAYS OF RECEIPT ALONG WITH PROOF OF PAYMENT.

Licensor retains the ownership of this copy and any subsequent copies of the Software. This copy is licensed to you for use under the following conditions:

Permitted Uses. You may: use the Software on any supported computer configuration, provided the Software is sued on only one such computer and by one user at a time; permanently transfer the Software and its documentation to another user, provided you retain no copies and the recipient agrees to the terms of this Agreement.

Prohibited Uses. You may not: transfer, distribute, rent, sublicense, or lease the Software or documentation, except as provided herein; alter, modify, or adapt the Software or documentation, or portions thereof including, but not limited to, translation, decompiling, disassembling, or creating derivative works; make copies of the documentation, the Software, or portions thereof; export the Software.

LIMITED WARRANTY, DISCLAIMER OF WARRANTY

Licensor warrants that the optical media on which the Software is distributed is free from defects in materials and workmanship. Licensor will replace defective media at no charge, provided you return the defective media with dated proof of payment to Licensor within ninety (90) days of the date of receipt. This is your sole and exclusive remedy for any breach of warranty. EXCEPT AS SPECIFICALLY PROVIDED ABOVE, THE SOFTWARE IS PROVIDED ON AN "AS IS" BASIS. LICENSOR, THE AUTHOR, THE SOFTWARE DEVELOPERS, PROMPT PUBLICATIONS, HOWARD W. SAMS & COMPANY, AND BELL ATLANTIC MAKE NO WARRANTY OR REPRESENTATION, EITHER EXPRESS OR IMPLIED, WITH RESPECT TO THE SOFTWARE, INCLUDING ITS QUALITY, ACCURACY, PERFORMANCE, MERCHANTABILITY, OR FITNESS FOR A PARTICULAR PURPOSE. IN NO EVENT WILL LICENSOR, THE AUTHOR, THE SOFTWARE DEVELOPERS, PROMPT PUBLICATIONS, HOWARD W. SAMS & COMPANY, AND BELL ATLANTIC BE LIABLE FOR DIRECT, INDIRECT, SPECIAL, INCIDENTAL, OR CONSEQUENTIAL DAMAGES (INCLUDING BUT IS NOT LIMITED TO, INTERRUPTION OF SERVICE, LOSS OF DATA, LOSS OF CLASSROOM TIME, LOSS OF CONSULTING TIME) OR LOST PROFITS ARISING OUT OF THE USE OR INABILITY TO USE THE SOFTWARE OR DOCUMENTATION, EVEN IF ADVISED OF THE POSSIBILITY OF SUCH DAMAGES. IN NO CASE SHALL LIABILITY EXCEED THE AMOUNT OF THE FEE PAID. THE WARRANTY AND REMEDIES SET FORTH ABOVE ARE EXCLUSIVE AND IN LIEU OF ALL OTHERS, ORAL OR WRITTEN, EXPRESSED OR IMPLIED. Some states do not allow the exclusion or limitation of implied warranties or limitation of liability for incidental or consequential damages, so that the above limitation or exclusion may not apply to you.

GENERAL:

Licensor retains all rights, not expressly granted herein. This Software is copyrighted; nothing in this Agreement constitutes a waiver of Licensor's rights under United States copyright law. This License is nonexclusive. This License and your right to use the Software automatically terminate without notice from Licensor if you fail to Comply with any provision of this Agreement. This Agreement is governed by the laws of the State of Indiana.